Neighborhood Branding, Identity and Tourism

This book delves into neighborhood branding by looking at the City of Orlando and the identities that set each neighborhood apart from others.

Orlando is an international tourism capital, known for its abundant theme parks that allow for an escape from reality. The word "Orlando" is almost synonymous with Disney and Mickey Mouse – and for good reason. This place's brand identity is so strong that outsiders have trouble realizing locals often have a drastically different view of the city. But what else is there? What other brand identities does the place have? The stories from this case study highlight how local stakeholders play a vital role in the success of an overall place brand while also taking steps to maintain their own unique neighborhood vibes.

This book will be valuable to academics and students interested in neighborhood branding and shaping identity from the perspective of tourism, geography, and urban studies.

Staci M. Zavattaro, PhD, is associate professor of public administration at the University of Central Florida. She serves as editor in chief of *Administrative Theory & Praxis* and focuses her research on place branding, administrative theory, and social media use in government.

Routledge Focus in Tourism

The *Routledge Focus in Tourism Series* offers a forum for cutting-edge research on a wide range of topics and issues within tourism studies. The series provides a forum for short topics aimed at specialized audiences and in-depth case studies that draw on a particular geographic locale. The format for the series is distinctive: each Focus is longer than a journal article and shorter than a traditional monograph.

Entertainment Tourism
Jian Ming Luo and Chi Fung Lam

Neighborhood Branding, Identity and Tourism
Staci M. Zavattaro

For more information about this series, please visit: www.routledge. com/Routledge-Focus-in-Tourism/book-series/FT

Neighborhood Branding, Identity and Tourism

Staci M. Zavattaro

Routledge
Taylor & Francis Group

LONDON AND NEW YORK

First published 2019 by Routledge

2 Park Square, Milton Park, Abingdon, Oxfordshire OX14 4RN

52 Vanderbilt Avenue, New York, NY 10017

Routledge is an imprint of the Taylor & Francis Group, an informa business

First issued in paperback 2020

British Library Cataloguing-in-Publication Data
A catalogue record for this book is available from the British Library

Library of Congress Cataloging-in-Publication Data
A catalog record has been requested for this book

ISBN: 978-1-138-57356-7 (hbk)
ISBN: 978-0-367-67187-7 (pbk)

Typeset in Times New Roman
by codeMantra

Contents

Figures

Acknowledgments

I would like to thank everyone who took time to speak with me for this project. I thank the generous City of Orlando staff members who explained neighborhood relations to me in detail, along with some of the city's interesting history. The economic development professionals to whom I spoke clearly have a passion for their work, which they expand to not only monetary success but community success as well. To the residents of Orlando, thank you for educating me about your neighborhood, its strengths, its weaknesses, and your wishes for a strong community.

I used their words when possible, but any mistakes herein are my own.

Terms and neighborhoods

Audubon Park – Neighborhood in the northern part of Orlando on part of the former Naval Training Center. Known for organic, sustainable businesses and shopping. Homes are older and sell quickly.

Baldwin Park – Planned development using new urbanism principles on the northern end of Orlando. Characterized by single-family homes, a mixed-use village center area that includes shopping and dining, multifamily homes, and Lake Baldwin.

Celebration – Separate town nearby Orlando known as the town Disney built. Included in this book because of proximity to Orlando and strong identity. Characterized by new urbanism features, along with a shopping and dining district on a lake.

College Park – Older neighborhood in the northwest portion of Orlando. Unique feature is streets named after colleges and universities (though there is not one in the neighborhood). Smaller, bungalow-style homes and shopping define the area.

Communications and Neighborhood Relations (CNR) – City of Orlando department responsible for internal and external communications (websites, press releases, social media, etc.) and neighborhood outreach.

Lake Nona – One of Orlando's newest neighborhoods in the southeast near Orlando International Airport. Characterized by Medical City, recreation, a trail system, and shopping and dining. Laureate Park is a main neighborhood in Lake Nona.

Mills 50 – Main Street District with three neighborhoods in the northern part of Orlando. Close to Baldwin Park and Audubon Park. Characterized by hip shopping and dining. Known as a trendy, up-and-coming area.

Thornton Park – Neighborhood near downtown Orlando and close to Lake Eola. Characterized by older homes and trendy bars and restaurants.

Introduction

Orlando is very much not Disney. Disney is a whole other city. Literally a whole other city. It makes me upset when people say, "I don't want to come to Orlando because I don't want to come to Disney."
– Orlando native and Thornton Park resident

I grew up a Florida kid. My family moved from New York when I was 2, so I am almost a native. Almost. I remember going to Walt Disney World for the first time as a child. For us Florida kids, the trip feels like a rite of passage. You never forget the first time you meet Mickey Mouse and Minnie Mouse in person. (I should note that this is really scary for some children because the characters are much bigger in person than on television or in books. Crying happens. A lot. There is a photo of me crying the first time I met Bert and Ernie from *Sesame Street*.) My parents surprised me with a trip to Disney when I was about six years old. We saw Cinderella's Castle, a luau at our hotel, and Main Street USA in Magic Kingdom. From then, I was hooked. I celebrated my 30th birthday at Magic Kingdom. I joked (sort of) that when I moved to Orlando in 2015 for my job, I would buy an annual park pass before getting real furniture. And that is exactly what happened.

Having been raised in South Florida, I assumed I knew what to expect when I moved to Central Florida. Orlando, as it turns out, is quite different than my hometown three hours south. It is big. It is bustling. It is, as almost everyone knows, home to some of the world's most sought-after tourist destinations, such as Walt Disney World Resort (this includes the Magic Kingdom, Epcot, Animal Kingdom, and Hollywood Studios theme parks), Universal Orlando Resort (this includes the Universal Studios and Islands of Adventure theme parks), SeaWorld, and others. In 2016, more than 68 million people visited

Orlando and its surrounding cities in Central Florida (Pedicini, 2017), and 2017 saw more than 72 million visitors to the area.

But as the Orlando resident quoted earlier points out, there is so much more to the city than those tourism destinations. Granted, much of the region's economic churn is thanks to those theme parks, but the local economic development partnership's branding campaign is "Orlando – You Don't Know the Half of It" (Orlando Economic Partnership, 2018). The campaign focuses on the growing technology presence in the city and region, along with the medical city presence near the new Lake Nona neighborhood. The Economic Development Partnership touts the city's talent pipeline thanks to many nearby colleges and universities, infrastructure, housing options (though affordable housing is a problem in Orlando, as it is in other cities), and quality of life as some of the reasons it is becoming a top economic destination for businesses (Orlando Economic Partnership, 2018). These are often areas and aspects of the city that the millions of visitors do not see when they head straight to the theme parks.

And that is okay. Orlando has a story to tell. I usually study city branding efforts, wanting to better understand how local governments develop, communicate, and evaluate strategic branding and marketing campaigns. Looking around Orlando gave me an idea to really examine a more micro level, a level sometimes more personal to us: the neighborhood. Driving through Orlando, which is quite spread out, one can immediately see the neighborhood identities. Sometimes, physical signage points you to a neighborhood. Sometimes, the housing style is the main identifier. Shops and restaurants, too, give neighborhoods character. I began to wonder what shapes a neighborhood brand. What gives a neighborhood its identity? Why do people choose to live where they do? What does Orlando do to foster neighborhood health?

To answer those questions, I asked people why they live in Orlando and specifically what made them choose their neighborhood. I spoke to long-time residents, newcomers, city officials, Realtors, and economic development professionals to gain a better understanding of Orlando's neighborhood philosophy. What you will read here is the analysis of those interviews. Taken together, the findings show what makes up a neighborhood brand, identity, and image.

The idea for this book came to me as most good ideas do – on the verge of sleep. I wanted to find a new path for my research, and living in Orlando has showed me that neighborhood identity can be both unifying and divisive – sometimes simultaneously. I wanted to better understand why people decide to live in certain neighborhoods. Do the pragmatic reasons take center stage? How do emotions play into the decision? For me, I chose my neighborhood for both kinds of reasons. I live in an area called Baldwin Park, which usually elicits an "oooh" from people whom I tell. Baldwin Park is a planned community in the city, built where an old U.S. Naval training facility once stood. Longtime residents remember when this was an empty plot of land that the city purchased for quite the steal, according to one resident who has lived in and worked for the City of Orlando for nearly 40 years.

Baldwin Park is known for its large homes, some starting at least one million dollars. I, however, live in a one-bedroom apartment, which I often feel compelled to tell people. It is as if I am telling them, "I am not *that* person." But am I? I chose this neighborhood because it was far enough away from campus to not see students all the time. It is safe (according to statistics and my feelings of safety). I like the idea of sidewalks, given that the state I lived in before coming back to Florida saw sidewalks as a big government intrusion so they were few and far between.

I realized I chose this neighborhood for both pragmatic (not too far from work; close to restaurants) and emotional (safe; walkable) reasons. I wondered if others did the same. The goal of this book is to find out how, if at all, neighborhood brands, identities, and images influence a person's decision to live in a certain Orlando neighborhood. Such an endeavor is needed as there are few studies about neighborhood branding, even though it is taking place more and more (Johansson & Cornebise, 2010).

Methodology

To answer the research questions, I used qualitative interviews of stakeholders throughout Orlando. I interviewed city officials, economic development officers, local Realtors, and those living in the city. Within the city, key information came from the Communications and Neighborhood Relations (CNR) department, other city officials, and a city historian. I used snowball sampling, asking people to refer me to others who might be willing to speak with me about this project. I recorded each interview in line with my university's Institutional Review Board procedures and the person's permission. I then transcribed the interviews and analyzed each for patterns. Thirty people contributed to the interviews from Orlando. To be honest, finding people to speak with me was more challenging than anticipated. People would agree to an interview, then never follow up when I reached out to set times. People also would not respond to my requests for interviews.

Each interview lasted approximately one hour. I analyzed the data for patterns, and the chapters reflect those findings. I did an inductive coding (Berg, 2001) because the project is exploratory in nature, and I am not testing hypotheses. For instance, I coded themes such as community, family, and safety as emotional reasons for choosing a neighborhood. Pragmatic reasons for choosing a neighborhood included being close to schools, shopping, parks, and more. One interviewer told me he picked a neighborhood with his family because "There were activities in the community I felt the girls (his children) would benefit from, and we were still central enough to go other places and do things." These are pragmatic reasons for choosing a community.

Interviews were appropriate because this is an exploratory study as there are few definitions of a neighborhood brand. All told, each theme contributes to an overall picture of neighborhood branding, identity, and image – all terms seemingly taken for granted in the nascent literature in this area. Readers will see there are no names used in the book, and this is in line with Institutional Review Board procedures. While some interviewees are public officials whose names I could use, I chose not to for continuity throughout.

Plan of this book

As mentioned, the book is organized thematically based on analysis of interview transcripts. Chapter 1 asks the question "what is a neighborhood?" to set the foundation for the project. Included in that chapter also is literature on place branding, image, identity, and place attachment to set the theoretical foundations. My background is in public administration – hence the study in this book of neighborhoods as a mechanism to interact with government officials while also creating a sense of shared neighborhood governance and pride. In my field, public branding is becoming a core governance strategy (Eshuis & Klijn, 2012) that deserves increased attention. Neighborhoods seemed like a logical place to extend this line of inquiry, which usually focuses on nations, states, and cities, given neighborhoods are often deeply personal. Sometimes, neighborhoods are so personal that a person's identity becomes intertwined with the place's (Tyson, 2014).

Chapter 2 gives a short history of Orlando, along with the City of Orlando's neighborhood relations philosophy. The city has, of course, grown substantially since its founding in 1875, when agricultural ventures, specifically citrus, drove economic progress. Today, tourism is a leading industry in Orlando, generating $255 million from hotel taxes in Orange County for 2017 (Russon, 2018). Tourism takes a toll on local infrastructure and sometimes causes a divide (Bastias-Perez & Var, 1995), as the person quoted earlier indicated. City and county employees need to cater to everyone yet take a strong focus on building strong neighborhoods. It is strong neighborhoods, the CNR team explained to me, that create a vibrant, active community. The chapter details their neighborhood relations strategy and how they help areas create a strong sense of identity.

In Chapter 3, the emotional reasons for selecting a neighborhood are detailed. I coded these as emotional reasons because they seem difficult to measure and quite individual to respondents. The three main themes to emerge from the interview coding are sense of community, feelings of safety, and new urbanism. Each is explained in detail within the chapter, and they are important components of creating neighborhood identity. Emotional attachment to a neighborhood can help or hinder feelings of attachment and belonging (Greenberg, 1999), and thus affect an overall identity. If a neighborhood has a negative image, then people try to avoid living there. Sometimes, people even avoid driving through so-called bad neighborhoods (Dilulio, 1996; Stark, 1987), so the emotional reasons become important to understand.

Chapter 4 details pragmatic reasons people choose a neighborhood, and by pragmatic, I simply mean practical rather than the philosophical definitions. Reasons included green space, close to work, good schools, and recreation. Schools are a major reason for which parents choose a neighborhood (Barwick, 2014), as are parks and recreation opportunities, especially in high-density development (Morrow-Jones, Irwin & Roe, 2004). Those in this study cited many of the same characteristics when it came to neighborhood choice and identity.

In Chapter 5, I detour a bit to discuss the Pulse nightclub shooting. On June 12, 2016, a lone gunman murdered 49 individuals at the LGBTQ nightclub Pulse on Latin Night. I included the incident in this book because it changed Orlando's identity forever, for many reasons. The chapter is not about the first responder efforts but reflects the overall city and community cohesion after Pulse. Many groups came together to support one another, and there are still signs of remembrance around the Orlando community. The #OrlandoUnited moniker is still quite present in the city, as are rainbow colors on murals and other art ventures.

Chapter 6 concludes the book by offering a definition of neighborhood identity and branding. I save this for the end, given that the elements throughout combine to create a working definition others can use when studying this concept. There are a handful of studies that look at neighborhood branding (Wherry, 2011), but they rarely define the term. So, my hope is that a working definition can be shaped and modified through time. I also want to give a brief note on what this book is not. My previous research has focused on city branding and marketing. I wanted to explore something different based on seeing the neighborhoods in Orlando seemingly have their own unique identities. So, this book is not about city branding, although many people to whom I spoke thought as much. I would tell people about this research, and their reaction would be something like "Have you talked to someone from City of Such and Such? They really revitalized their downtown area, and now the city really has a strong brand." There is, of course, a connection between city branding and neighborhood branding – strong cities will likely have some strong neighborhoods – but cities were not my focus here. The interest is in how neighborhoods form an identity and why that matters, if at all. The city examples people mentioned (Cocoa Village in Cocoa Beach, Florida, was one, and Delray Beach, Florida, was another) are all strong contenders for how economic development and place branding work hand in hand, but those examples are not my focus herein. Instead, I asked people to whom I spoke about their neighborhood, why neighborhood branding and identity matters (again, if at all), and why they chose their neighborhood.

1 So, what is a neighborhood?

Orlando is built on the peel of an orange.
                                    – South Florida Railroad pamphlet

As the Orlando resident quoted in the Introduction aptly explained, the city is synonymous with tourism, especially theme parks. In 2017, more than 72 million visitors came through Orlando, becoming the first U.S. destination to surpass 70 million visitors (Shelton, 2018). In a press release announcing the statistics, Visit Orlando – the City's official tourism branding association – lauded all the new theme park rides opening at Disney World, Universal Studios, LEGOLAND, and SeaWorld. The release also mentioned tourist hotspots such as the ICON (essentially a large Ferris wheel akin to the London Eye), the Orlando City professional soccer team, and the new Lake Nona neighborhood developments (Shelton, 2018). It is no surprise, then, that Orlando's global brand image is strong (Fallon & Schofield, 2004).

But what else is there to Orlando? I thought the answer could come in its unique neighborhoods so set out to explore the concepts of neighborhood identity, image, and branding through the Orlando case study. As this chapter explains, there is little yet known about neighborhood branding, given much of the place branding research focuses on countries, states, and cities. Neighborhoods, whether in Orlando or Amsterdam or Sydney, have personalities of their own, for good or ill.

Why study neighborhoods?

> There is power in the idea of the neighborhood, power that comes not from its precision as a sociological construction but from its nuanced complexity as a vernacular term. Neighborhood is known, if not understood, and in any given case, there is likely to

be wide agreement on its existence, if not its parameters. Unfortunately, this generalized notion of neighborhood is not very useful in in- forming policy or planning for social change.

(Chaskin, 1997, p. 523)

In other words, neighborhoods are important units of study because of the inherent ambiguity. People make up the neighborhood, so that was my task in asking people their why – why there?

The remainder of the chapter highlights the research on neighborhoods, trying to offer a common ground based on existing literature. Next, I define neighborhood identity, image, and brand. Not surprisingly, there are myriad definitions of all the terms, especially when it comes to understanding what exactly a neighborhood is. Finally, I conclude with literature on place attachment, as it is germane to the remainder of this project. Place attachment helps explain why people select, or deselect, a place such as a neighborhood.

But what is a neighborhood?

This question seems like an easy one, but there is little agreement even among experts (Campbell et al., 2009). "Indeed, 'neighborhood' is a vague, difficult-to-define, concept. Scholars investigating the significance of neighborhood for individuals' behavior and well-being often do not provide the term with an explicit definition" (Gou & Bhat, 2007, p. 32). Campbell et al. (2009) explain that neighborhoods not only include physical boundaries but also are created through neighbors' spatial interactions – or lack thereof. As such, more precision is needed when it comes to defining neighborhood, especially given the geo-boundary constraints inherent in neighborhood design and conceptualization (Siordia & Seanz, 2013). Typically, a neighborhood is thought about as a defined geographical space with dwelling units and people linked together through social bonds. Chaskin (1998) notes a neighborhood can be a spatial unit, a social unit, and a relational network. These are not mutually exclusive so make the study of neighborhoods both interesting and challenging.

Typically, the U.S. Census tract is used to define a neighborhood's physical boundaries as a Census tract is a

small, relatively permanent statistical subdivisions of a county or county equivalent and generally have a population size between 1,200 and 8,000 people, with an optimum size of 4,000 people. The Census Bureau created census tracts to provide a stable set of boundaries for statistical comparison from census to census.

(U.S. Census Bureau, 2018, para. 1)

Yet studies using only these tracts often leave out what people perceive as their neighborhood boundaries (Coulton et al., 2001), given those perceptions might differ from official geographical boundaries. Chaskin (1997) is surer of his definition of neighborhood, writing a neighborhood is *"clearly* a spatial construction denoting a geographical unit in which residents share proximity and the circumstances that come with it" (pp. 522–523, emphasis added). Given the complexity in the definition, Gou and Bhat (2007) argue that we should measure neighborhoods by what matters to the people in them. That is the task of this book. I am less interested in the physical boundaries of a neighborhood and more in how people perceive their neighborhood identities or those identities found in other neighborhoods.

Synthesizing the literature, Gou and Bhat (2007) find that neighborhoods seem to share some similar characteristics. First, neighborhoods are *geographically bound*, but meaning depends on the function and domain of the space. For instance, people should take into account what is being done and studied in the neighborhood rather than generalizing that something happening in one section of a neighborhood applies to another. (To illustrate what Gou and Bhat (2007) mean by neighborhood sections, where I live, in Baldwin Park, there are ten distinct neighborhood regions within the larger geographical area that is Baldwin Park. People who live closer to the Winter Park side might have a different experience than those living closer to the city center, which is on the other side of the neighborhood.) Second, neighborhoods have both *fixed and subjective elements* (Guo & Bhat, 2007). A roadway or lake might be fixed, but how people use and interpret those features is up for debate. And third, oftentimes *administrative definitions* of neighborhoods (such as Census tracts or city maps) are imperfect but offer a start for studies and government operations.

Neighborhoods could promote or hinder citizen engagement, depending upon factors such as size, age, and type of housing (Haeberle, 1987). Income, at least in Haeberle's (1987) study, is not a good predictor of neighborhood engagement. In Orlando, for example, the Parramore community is a lower economic-status neighborhood but recently gave rise to one of the city's most active neighborhood associations. Resident Vinny Carter reignited the Carter Street Neighborhood Association after being fed up with crime in her neighborhood. There is a push to make the neighborhood safer and more engaged thanks to Carter's efforts in the community she has called home for more than 60 years (Giorgio, 2017).

Frankly, no question gave people to whom I spoke more pause than this one: what is a neighborhood? People literally took pauses to think or said things like "hmm there's a good question" or "wow, what is a neighborhood?" I did not expect that, but it makes sense and aligns with existing research on the topic (Gou & Bhat, 2007). Neighborhoods are often the unit of analysis in many studies, but the term seems to be taken for granted. I do not pretend to offer a definition, as that was not my goal. My goal was to better understand neighborhood and identity and branding through the people who live in Orlando's neighborhoods. While the results are germane to Orlando, other communities can learn lessons by asking their residents about branding and identity at this micro level.

When asked the question "What is a neighborhood?", people responded in various ways. Some examples include:

Hmm, that's a tough question. A neighborhood is obviously a place where people live so they sleep there at night. But a neighborhood is also a community of people who can ban together and help people out when they need it.

– Audubon Park resident

A neighborhood is an encompassing community that brings the environmental, people, social, and intellectual together, hopefully dependent upon the community for the good of those who live there. There's a level of cooperation and collaboration, reliant upon what people have to contribute.

– Baldwin Park resident

So a neighborhood is numerous homes clustered together with a sense of friendliness and collegiately, camaraderie among the people who live there. To me a neighborhood is safe, and comfortable, and serene.

– Celebration resident

I think a neighborhood is just a group of people that basically have the same socioeconomic backgrounds. You get into most neighborhoods, your age range is usually within 10 years. Most of them you have your younger neighborhoods that have small kids [and] are usually built around schools now.

– retired City of Orlando employee

Boy there's a question. A lot of it is just geography, its identity within that geography. In my experience, neighborhoods should have a central point where everybody knows to converge during certain times of the year.

– Main Street director

These are just a sampling of answers to the question and some commonalties that carry throughout this book. First, one sees that respondents equate neighborhoods with people. This is where a sense of community becomes important for neighborhood identity. Some in this study bemoaned not knowing their neighbors and missing that sense of community. One local Realtor told me her neighbors keep their blinds closed, or people will pull cars directly into their garage to avoid saying hello to others. Second, there is a connection to place. Pragmatically, neighborhoods are geographic entities. They can be spatially measured and assessed. This is useful for many reasons, including but of course not limited to taxing, policy, and public service provision. Emergency management, too, can tell a lot about neighborhoods if they are in, say, a flood zone. Third, neighborhoods are about emotional ties such as safety, connection, and help. Whereas you can measure plots and land in a geographical spatial sense, these more emotive reasons are more difficult to capture. How do you measure nice, for instance? These qualities are no less important, though, when it comes to creating a neighborhood brand and identity.

Neighborhood identity, image, and branding

Brand identity in general is understood as how the organization creates and presents its brand (de Chernatony, 1999). The point is the brand identity originates from the organization itself (Nandan, 2004). Organizations then communicate brand identity through various marketing strategies. Related, brand image is how people perceive and understand the brand identity (Nandan, 2004). That is where brand subjectivity comes in, given people interpret the brand through their own lenses. We probably all know someone who loves a certain product brand or destination brand while having equally negative feelings about another. That is the embodiment of brand image and brand loyalty.

Trying to define a neighborhood brand is difficult as best, so I start first with what is a place brand. Not surprisingly, there is no agreed-upon definition of a place brand. Put simply, place branding refers to developing and communicating brands for certain geographic regions, ideally to create positive associations in peoples' minds (Eshuis, Klijn & Braun, 2014). Related, place marketing is the tools used to communicate that brand identity, such as a website, press release, or social media. Branding and marketing are active strategies used to define and promote a place, while the brand is the associations between the communicated identity and realized image.

Relating brand image to neighborhoods, each year the American Planning Association (APA) puts out a list of "America's Best" including neighborhoods, streets, and public spaces. Talen, Menozzi, and Schaefer (2015) analyze potential conflicts between the "great" neighborhoods and social equity. In their article, the authors detail some of the APA criteria, including neighborhood characteristics (social activities, transportation, architecture, character), neighborhood form and location, personality, and sustainability, for instance. The APA selects from self-nominated neighborhoods, so the neighborhoods selected often present an idealized version of American neighborhoods (Talen et al., 2015). Granted, looking at the APA website listing for the great neighborhoods, there is diversity in rural versus urban, but all have elements of new urbanism, such as walkability, bikeability, sense of community, and complete streets. But Talen et al. (2015) found that many of the great neighborhoods are gentrifying and not accessible to those who are economically disadvantaged. "It is also a call to come to grips with the reality that America's best-loved neighborhoods may be increasingly out of reach to anyone but the affluent" (p. 131).

What is most interesting about the great neighborhoods, streets, and public places is they are often ideal versions of what people have in their minds when they think of a neighborhood. The ideal is a key part of why people feel connected or disconnected to their neighborhoods. When I asked people their ideal neighborhood for this research, many mentioned walkability, bikeability, recreation, access to stores and restaurants, and safety. These kinds of neighborhoods do exist in practice but might leave out other conceptualizations of neighborhood living. Together, though, these great neighborhoods show the importance of identity and neighborhood branding.

There are few studies that explicitly look and neighborhood branding strategies. Wherry (2011) details neighborhood branding efforts in a Philadelphia barrio neighborhood called Centro de Oro. Wherry (2011) met and interviewed people he called cultural entrepreneurs who charged themselves with branding the neighborhood for social and economic development. As he explains, neighborhood branding is complex and includes many components, people, and activities. Branding activities can be both intentional and unintentional, by purposefully branding the neighborhood through logos, slogans, and events, to people becoming brand ambassadors promoting the neighborhood wherever they go. For Wherry (2011, p. 4),

in the context of a neighborhood, the brand is apparent from what businesses sell, how their storefronts are designed, what kind of music emanates from open neighborhood widows and passing cars, and what kinds of themes are depicted on the neighborhood's plentiful murals.

Wherry (2011) studied a neighborhood that was stigmatized for its Latino culture and depressed economic status. A challenge, then, for purposeful neighborhood branding was to remain authentic without selling out. Sometimes, though, the narratives against the truth of the neighborhood became so powerful they drowned out negative perceptions. The problem, then, is the disconnection between promises and reality. There is a risk of exploiting ethnic neighborhoods, and "good" marketing could drive out existing authenticity through gentrification (Wherry, 2011).

In their work, Masuda and Bookman (2018, p. 166) write:

> neighbourhood branding refers to the symbolic and material practices of state and/or private cultural producers who aim to enhance the appeal of local areas within the city in order to attract investment, promote consumption, reduce criminality, or to achieve social and cultural aims such as invoking civic pride.

Neighborhood branding primarily emerged as a tool for economic revitalization, but locals can join together to engage in counter-branding strategies in an attempt to preserve their right to the city and place (Masuda & Bookman, 2018). Given the hyper-local level, neighborhood branding often is more politically charged and contested than city or nation branding (Masuda & Bookman, 2018). At this level, branding strategies go beyond only economic outcomes to include social and political outcomes as well (such as feelings of safety, sense of community). As a result, "neighbourhood branding is entangled in and co-configurative of urban cultures and notions of place via a range of activity conducted by a host of actors with varying and often conflicting interests" (p. 171).

Similar to Wherry (2011), Johansson and Cornbiese (2010) also examine an ethnic neighborhood in Chicago for their neighborhood branding case study. They look at Andersonville, a predominantly Swedish enclave that is not an officially designated neighborhood by the city. They argue that neighborhoods take to marketing and branding as a way to build community while also potentially countering (or even embracing) market forces changing the neighborhood. They detail how brand managers and ambassadors came together to create events and celebrations that highlight Swedish culture. Realtors and neighborhood associations are powerful, shaping what the neighborhood wants to be (Johansson & Cornbiese, 2010). A challenge, though, is that Swedes are leaving the neighborhood, leaving the identity in question so leaders are exploring newer branding strategies.

In his study of Abbot Kinney Boulevard in Venice, California, Deener (2007) shows how through time the street, lined with local businesses, became a physical and psychological dividing line for the neighborhood brand. Based on interviews with shop owners and residents of the neighborhood, Deener (2007) found the street is a racial dividing line and serves as a source of nostalgia for many who lived near or along the street since youth. But for newer residents, the street was a way to reframe neighborhood identity. As Deener (2007) indicates, the street was renamed from West Washington Boulevard to Abbot Kinney after the area's original developer to give the street new life. Indeed, they marketed it as "the newest street in Los Angeles" (p. 298).

They also created a neighborhood association made up of business owners, and membership fees went toward street beautification projects such as benches and bike racks. The coalition tried to create identity by adopting festivals and a Bohemian feel along the street but also erased notions of gang violence that pervaded the area in the 1990s. Despite trying to erase the issue, racial division still is present along the street.

> Newcomers make efforts to preserve the street's commercial culture, believing the anticorporate branding is central to Venice's bohemian identity, but they do not make equal efforts to protect the use of the street for long-time residents who cannot afford to shop at the new upscale commerce.
>
> (Deener, 2007, p. 293)

Using neighborhood reputation as the foundation for their study, Pais et al. (2014) explore neighborhood reputation in the wake of the housing crisis in the U.S., specifically the City of Las Vegas. They argue that neighborhoods with positive reputations are critical to creating overall healthy cities, and the relationship also works in the reverse. What is important to both, they note, is resident perceptions of neighborhood reputation. Findings indicate about an even split regarding resident perceptions of neighborhood reputation even after the foreclosure crisis hit hard. The mitigating effect was an already positive neighborhood reputation; in other words, if residents felt their neighborhood was strong, the effects of the crisis were lessened (Pais et al., 2014).

Neighborhood branding, though, is tricky, as residents might not understand or see the direct effects. In their study, Rich and Tsitsos (2016) examine an arts district in Baltimore, Maryland, called Station North Arts & Entertainment District. After analyzing interviews with stakeholders, they found that many residents perceive the district only as a branding trick that will have little or no economic benefit for them. For example, usually arts districts come with promised tax breaks that never materialize, causing frustration among residents. Despite skepticism regarding the tangible benefits, residents do seem to buy into the positive identity-related aspects of the brand. For instance, the branded area might become more attractive for future residents and business owners who want to be associated with the arts district (Rich & Tsistos, 2016).

Only a handful of these studies operationalize neighborhood branding, though. Wherry (2011) and Masuda and Bookman (2018) by detailing specific strategies neighborhoods can use to build an overall brand identity. Branding is an active strategy to foster and disseminate a place identity. It is a process that ideally involves relevant stakeholder groups in a meaningful way. NeighborWorks America is a nonprofit organization focused on creating healthy communities through neighborhood partnerships. In their documentation explaining neighborhood marketing and branding, NeighborWorks America define a neighborhood brand simply as "what people think of your neighborhood" (Kelsh, 2015, p. 7). Relatedly, Farris and Kendrick (2010) define a neighborhood brand as "recognition among residents that a particular business excels delivering a product, selection, service, or experience" (p. 6). They rely on a more business-minded definition usually reserved for products. They do, though, say that neighborhood brands are composites of stories that are on tops of peoples' minds. In other words, a neighborhood brand is socially constructed.

I asked the City's CNR team why a neighborhood would want an identity and brand. As one member explained, "It goes back to that particular neighborhood having pride, that history. When I got out to certain neighborhoods, I ask them 'tell me who you are.'" She explained that as Orlando continues to grow rapidly, more people are looking for a connection to their neighborhoods through history and a strong identity. There is such a disparity with Orlando neighborhoods, given some are older and historic and there are new areas such as Lake Nona. As she said, Lake Nona is developing an identity from the ground up, while other residents throughout the city are fighting to preserve historical identities as builders tear down small bungalow homes and replace them with two-story modern marvels. "Downtown neighborhoods are feeling the squeeze of development," she said.

The CNR assistant director said that the city feels strongly about strengthening and protecting its neighborhoods and recognizes areas for improvement. When asked to define a neighborhood, she paused before saying, "I see it as a community where people live that on the borders are those personal services and other businesses they need on a regular basis to support their lifestyle. To me a neighborhood is a sense of place." But she recognizes there are many sides. From her view in her office, neighborhoods are about what residents want, while from an economic development perspective, the neighborhood takes on a different definition.

Place attachment

Given that there are few definitions of neighborhood branding, one can look at the environmental psychology and human geography literatures for additional guidance. A concept that can help with this study is place attachment, defined "as an affective bond or link between people and specific places" (Hidalgo & Hernandez, 2001, p. 274). As Hidalgo and Hernandez (2001) point out, many definitions of place attachment are quite broad and not easily distinguishable from measures of residential satisfaction. Many studies focus on neighborhood attachment, yet the approach is not without flaws. The neighborhood level is studied because an underlying normative assumption is that people are more emotionally invested in a neighborhood than in a city, region, or state (Hidalgo & Hernandez, 2001).

Place attachment is inherently linked with place identity, as there is an important relationship between physical space, emotional connections, and personal identity (Ujang, 2012). Symbolic aspects of place, too, are integral to creating attachment. For instance, in Orlando, when you tell people you live in Baldwin Park, an upscale neighborhood, it says something about you – whether you want it to or not. The same is said if you tell people you live in Pine Hills, a neighborhood on the nightly news for instances of crime. Perception is reality.

While the neighborhood is often the core unit of study, researchers have also examined attachment to places such as streets, cities, public spaces, and more (Lewicka, 2011). Neighborhood attachment is important to understand because positive attachment can lead to increased community engagement, while negative attachment could have the opposite effect (Comstock et al., 2010). Several factors can influence neighborhood attachment, including how long someone has lived there, if they own versus rent, physical conditions (broken windows, for instance), community benefits (events, community garden), race, ethnicity, and other socioeconomic factors (Comstock et al., 2010). The relationships are exacerbated at the block level, where people often feel more connected (or not) to immediate neighbors (Brown et al., 2003). This relationship is vital when a community faces deterioration, as the stalwarts might try to preserve that sense of identity (Brown et al., 2003).

Researchers often study the connection between place attachment and economic outcomes such as safety (Delisi & Regoli, 2000; Silver & Miller, 2004), neighborhood use value (Taylor, 1996), and personal health (Carpiano, 2008). These are economic outcomes because there is a cost-benefit associated with each. In neighborhoods people perceive as blighted, there is an economic disincentive to move there (see also Tiebout, 1956). Gentrification, then, could come into play bringing with it other social, political, and economic problems. Personal health also has an economic component when it comes to paying for illness. The economic impacts are important but should also be coupled with psychological, affective outcomes (Manzo & Perkins, 2006).

Given that, Bolan (1997) indicates that studies of neighborhood attachment often fall into two categories (not mutually exclusive): attitudes about the neighborhood and behaviors within the neighborhood. The two are connected given someone's attitude about the neighborhood influences how and why people connect and participate. The connectivity is crucial to explore given people (usually) find satisfaction via those social ties (Kasadra & Janowitz, 1974). Length of residence in the neighborhood and how people get to know each other influences the strength of those ties (Kasadra & Janowitz, 1974).

Conclusion

All told, place branding within public administration is becoming a new governance strategy worth understanding more about (Eshuis & Klijn, 2012). As such, we can begin exploring the different levels of places and connections. Typically, place branding research focuses on nations, states, and cities, but moving down to the neighborhood level gives us another avenue to explore in a place where people spend most of their time. As public administration professionals, we can learn about what makes people tick, so to speak. For the City of Orlando, the CNR professionals told me that their biggest struggle is finding neighborhood leaders to get involved. Perhaps getting to know better the barriers can help overcome some of the problems.

2 Orlando's neighborhood philosophy

We make sure we can touch residents in any possible way so we have a connection.

– City of Orlando CNR team member

It is no accident that the City of Orlando has its share of strong neighborhoods. City officials are the first to admit that things are not perfect; several neighborhoods in Orlando's vast geographic expanse could be improved upon. Economic decline often drives people away from certain areas, leading to negative external perceptions (brand images) of the place. City officials try to give all residents tools to create vibrant communities, but there is no perfect solution.

In this chapter, I detail interviews with city officials and local Main Street business district leaders to highlight the city's neighborhood philosophy. First, though, I give a brief history of development in the Orlando area to show how the city has evolved from its early days as a trading post and stop along a railway to a bustling city center. This is not meant to be an historical analysis of Orlando but a way to show the city's continued evolution.

Orlando: a history in brief

There are many books written about Orlando's history. Perhaps the most famous comes from E.H. Gore, a local newspaperman. His book *From Florida Sand to "The City Beautiful"* was published in 1949 and revised in 1951. It is often considered the definitive history of the area, so I recommend it for further reading. I draw on his observations here to share how Orlando transformed from a small village to what it is today (and what it continues to become).

There is some disagreement about how Orlando got its name. According to Gore (1951), the city's namesake is a soldier named Orlando Reeves who was guarding fellow troops against Seminole Indian attacks during the Seminole War. Indians staged a surprise evening attack, and Reeves, who alerted his fellow soldiers, was killed by arrows. The next day, soldiers buried Reeves under a tree near Lake Eola, the city's downtown lake, with the word "Orlando" on a wooden slab as his grave marker. According to Gore (1951), there is debate where he was buried, but nevertheless, early settlers adopted his name for the first post office and kept the name during incorporation in 1875.

Aaron Jernigan was likely the first settler in the area in 1842, and his family followed two years later, when they moved into the area's first log cabin (Gore, 1951). Wooded trails eventually gave way to paved streets in the latter 1800s, thanks to the removal of nearly 500 trees (Gore, 1951). Even Gore observed that in the area's early days, people wanted to live in a location that was aesthetically pleasing. "The residents of Orlando take great pride in their homes and you will find them from the bungalow type to the large, palatial homes, surrounded by beautiful green laws and ornamental trees and shrubbery" (p. 5)

On June 23, 1875, residents of the Village of Orlando met during a public meeting, whereby 29 qualified electors showed to vote on creating the City of Orlando pursuant to state rules (Bacon, 1975). On July 31 the same year, an election was held where 22 qualified electors came, approving the new City of Orlando. The boundaries were "one mile due east, one mile due west, one mile due north, one mile due south forming a square" around the courthouse (Bacon, 1975, p. 61). Indeed, a pamphlet promoting the city's features would laud Orlando as "a perfect square" (Sentinel Print, 1910, p. 3).

Since its incorporation, the city's economy began to grow. Much of the commerce centered within main core – the perfect square of downtown. Cattle was the first major industry (Bacon, 1975). The Jernigans, for instance, brought their herd of 700 cattle to the region. "The agricultural nature of early Orlando made it necessary for homes to be built far apart. Visiting among neighbors was difficult and possible only for such occasions as holidays or church meetings held in private homes" (Bacon, 1975, p. 20). Early settlements, then, were not what we would think of as contemporary neighborhoods today, but there was indeed a sense of community during these gatherings, where people would sing and dance into the evening (Bacon, 1975).

According to a city historian to whom I spoke, Orlando grew because it was a respite from the evolving industrial centers in the U.S. Those cities became crowded and dirty quite rapidly, so Orlando, with its broad stretches of land and citrus groves, became a place to escape. The railroad nearby helped that growth, though it was surely an ordeal to get from Sanford where the train stopped to Orlando many miles away. He said:

> Before that you had to take a boat to Sanford and ride a horse or walk or carriage or something on a dirt path. It wasn't like we had roads. In general it was a half or day or more trip with a horse or cart. You'd have to take this circuitous little path that wound you here.

He explained the railroad cut the trip to about 45 minutes, something he called a "game changer" for the city.

For instance, in 1880, he said that the city's population (just five years after its incorporation) was about 200 people. In 1890, the population soared to 3,000 people thanks to the train. In Orlando, the train moved the city's core to Orange Avenue – away from its original, neat square blocks. Surrounding areas such as Sanford and Winter Park experienced both economic and population growth thanks to the railroads. "In all, 15 towns or settlements established between 1880 and 1895 on or very near the railroad line. The urban form, therefore, was one of small towns strung like beads along a string of railroad lines" (White, 1965, p. 17). By 1926, two distinct central cores formed as Central Florida continued to grow – Orlando and Winter Park. (Winter Park and Orlando today are neighboring cities.) Each had its own urban core and "the new scattered residential areas marked the first appearance of suburbia on the Central Florida map" (White, 1965, p. 26).

The city also was becoming home to one of the foremost citrus crops in the U.S. When settlers arrived in Orlando, they found native orange crops, likely leftover from prior failed Spanish takeovers of the land from Native Americans (Peroldo, 2017). A booming railroad system spurred the citrus industry growth, with farmers now able to export their crops far distances. A cold snap in 1894, though, threatened to destroy the industry. An alarm sounded, alerting farmers to cover their crops, but it was too little too late (White, 1965) and much of the crop turned black (Peroldo, 2017).

After the freeze, many farmers did not replant, so those areas turned into platted streets and lots. Many of the city's current historic districts, the city historian explained, were platted by the 1920s. Some locals wanted to rebuild the landscape, so they began replanting trees and other shrubbery, leading to the creation in 1908 of a city nickname that is still used today: The City Beautiful. A local contest led to finding the slogan (Gore, 1951). Because of the trees, lakes, and fresh air, people often made Orlando their winter home destination, but the favorable climate and land ripe for planting turned many of them into year-round residents. "As more people kept coming as a destination for a few months in the winter, you needed more people to actually live here to service all those industries that would result from that. It kept building on itself," the historian said.

What did he mean, though? Well, tourism has always been in Orlando's blood. The train brought people here to see swampland, escape harsh winters, and enjoy fresh air away from smoggy cities. "People would come down here because it was seen as a health benefit," he said. That included his grandmother who visited in 1905 with her father. She was either 16 or 17 at the time, he told me, and the two of them spent about six months in Orlando without their other family members. His grandmother decided to move to Orlando in the 1920s. She was a schoolteacher and moved with a one-year-old son, the historian's father. His dad grew up in Orlando, and the building where his grandmother taught still stands along Lake Eola today. "That was something that people did a lot of. You try to come down and see if you could make it down here," he said. Called the city's historic preservation guru, it is clear Orlando is in his blood.

I asked people at the city what, if any, was the first official neighborhood. There was no clear answer to the question, but the historian told me that the city grew from its center. The city's physical center, where it began, is now the Orange County History Museum in downtown. A plaque outside the museum indicates Orlando's origins were there. Homes were often further away given horse and buggies could take people to the clearer farmland areas. As Gore wrote (1951, p. 20), "the agricultural nature of early Orlando made it necessary for homes to be built far apart. Visiting among neighbors was difficult and possible only for such occasions as holidays or church meetings held in private homes."

Looking at Orlando today, you can see that growth pattern. The center has dense buildings, such as apartment homes, condo complexes, and large corporate offices. The suburbs follow traditional American development patterns, and you need a car to get almost anywhere if you live in the more suburban areas. One of the city's earliest neighborhoods formed around the Conway chain of lakes near the downtown core. In the 1850s, what is today the neighborhoods of Conway, Dover Shores, Belle Isle, and Azalea Park, the area became one of the first places to live outside of the Orlando core because of a paved Conway Road, one of Orange County's first paved highways (Candelaria, 2017).

The city historian told me that the downtown historic district is quite small, given many of the homes were destroyed in the city's early years. Looking at old Sanborn Maps, drawn by the Sanborn Fire Insurance Company, we can tell what kinds of homes once stood in the city's downtown core. The maps the city still has are quite old – dating to 1913 – and because they were all drawn by hand, the creators would simply erase or paste over existing properties. Looking at the old outlines, the historian showed me where homes had a Victorian style with round turrets that resembled castles. These features were drawn right onto the map. Probably the best examples of some of the city's oldest homes is in the current Lake Eola Heights neighborhood, with many of the homes dating to the 1880s and 1890s.

Early histories of Orlando tout the city's desirable qualities. For instance, one early history by the Sentinel Print noted that:

> That individuality which has always been the town's is as marked today as ever. Orlando is no suburb or outlying district of some other city, a condition that characterizes so many other towns in the country; it is never spoken of as being on the road that leads to some other place; it is its own business center.
>
> (Sentinel Print, 1910, p. 1)

Another pamphlet without an author encouraged people to visit or move to Orlando to enjoy amenities such as paved roads, polo clubs, active church life, pure water, shopping, banks, ice, gas, schools, and sanitariums (No Author, 1913). "It is a modern city, with paved streets, stone sidewalks, electric lights, gas, a complete system of sewage, and the purest of water" (No Author, 1913, p. 2). But, the writing warned, be prepared to work hard if relocating to Orlando, as the agriculture industry is not for one seeking to make a quick fortune.

One thing that changed the city's composition as it grew was highway development. This is a classic example of what happened in cities throughout the U.S. as the highway system came to fruition. Cutting through Orlando is Interstate 4, called I-4. I-4 runs from Tampa to Orlando and finished construction in 1965. Steady increases in tourism to Central Florida necessitated the highway's growth, but it was when Disney World opened in 1971 that the roadways really reached capacity. As of this writing, construction is going on what is dubbed I-4 Ultimate, a decadelong (at least) project to revamp a major portion of I-4 near the theme parks, all the way north to Altamonte (I-4 Ultimate, 2018).

There is a long history of highways causing social and geographic problems for neighborhoods. For instance, Baum-Snow (2007) found that each new highway built in the U.S. between the 1950s and 1990s caused an 18 percent population decline in urban cores. President Eisenhower's legislation led to this highway boom, making commuting throughout the U.S. easier, creating the family road trip. Highways also exacerbated political polarization by either physically dividing communities or causing the population transfer (Nall, 2015). Highways also aggravated racial disparities and segregation, often displacing or destroying traditionally black neighborhoods during construction (Bayor, 1988).

Orlando did not escape many of these problems with its highway construction. For example, the historian explained that the original Lake Ivanhoe was filled in to make way for the highway. Many neighborhoods, too, were divided. College Park, for instance, has the highway running through it. "It did lop things strangely," the historian told me. Downtown, too, was cut into pieces thanks to I-4. "When it came through downtown, a lot of this area, if you look at these old maps, it was a lot of warehouses. We don't have a warehouse district because of that."

By the 1920s, the city had an active planning department to both foster and regulate the rampant growth. Home building was increasing thanks to the train, agriculture industry, and tourism. "People were selling them and flipping them. It was a crazy time in the 1920s. There was a lot of speculation," the historian told me. While many people were building houses because they needed a place to live, others were hoping for the best. For example, the historian told me about a wealthy many who built eight Mediterranean-style homes of 2,000-square feet each to attract people to stay beyond winter. Each home came complete with its own grand piano. Some of the homes remain today on a street called Broadway Court.

He pointed out that boundaries are often political, and today, the city has decided that a neighborhood "has a clear boundary delineated by some main artery on the north, west, east, and south." As the city grew and expanded, he explained there was no rhyme or reason to creating neighborhoods until professional planning came into play. He said that people named plats after themselves, and one Florida state senator from the 1930s named Walter Rose platted areas himself, and neighborhoods with the name Rose in them are associated with his planning processes, explained the historian. Rose, though, was a Realtor and former president of the National Association of Realtors (National Association of Realtors, 2018).

Neighborhoods in Orlando are so varied and diverse. The city's downtown area is home to some of the city's oldest homes, but in the 1960s and 1970s, many of the homes were in disrepair, said the historian. "White flight had emptied downtown" because the area was seen as having high crime and drugs, he said. But eventually, people started buying and redoing the homes given the beautiful architecture. For example, people took a home that was once divided into several apartments and restored it back to one grand house. Said the historian:

> By having that strong sense of neighborhood identity, people felt comfortable they could spend money fixing up this house because they know it was going to be on the up and up if there were controls because of the historic district.

Historic districts, he explained, limit what people can do to homes in an effort to maintain their character and historic nature.

With neighborhoods growing and the city continuing to attract new residents and visitors, the neighborhood philosophy also has shifted. Indeed, the city has an entire team now dedicated to building strong neighborhoods, but there have been ebbs and flows.

City of Orlando neighborhood philosophy

As I noted at the outset, it is no accident that the City of Orlando is trying to build strong neighborhoods. Again, CNR team members admit that work remains and even lamented projects they wish they could have made happen but did not for whatever reason (money, time, resources, etc.). So, while they are trying each day to improve neighborhoods, they stress the secret to success is partnerships with local residents. As such, the team's main mission is to work closely with neighborhood association leaders – or create leaders in neighborhood to ideally create a strong identity.

The city's neighborhood philosophy began with Mayor Bill Frederick in about 1983, explained the department's assistant director who has been with the city for more than 30 years. When looking at the city map at the time, city staffers and the mayor noticed that it "looks like a great big jigsaw puzzle with pieces missing" because all the neighborhoods were color coded on the map. So, the mayor "decided he wanted to add a brand and a personality to the city, and he established the great neighborhood program, which evolved into neighborhood relations." Their initial task was to go out and put a face to all the city's neighborhoods, asking people what *they* thought was their neighborhood boundary. In addition to in-person meetings, the staff also did phone calls and mail surveys, "and that's how we came up with the neighborhoods map," she explained. Though the boundaries really have no legislative meaning, they are simply a way to organize areas of the city geographically.

For example, there is an area of Orlando called MetroWest. As of this writing, MetroWest has 112 different neighborhoods, said the coordinator, and 11 different homeowner's associations. Each asked to be included on the map, but the city said no because it was too complex. So instead, there is one designated area on the color coded map for what is geographically seen as MetroWest, encompassing 1,805 acres on the western side of the city (MetroWest Master Association, 2018).

The CNR assistant director explained to me the city follows an Asset Based Community Development (ABCD) approach in their office. The process involves community stakeholders leveraging community assets, and identifying potential assets, to create local economic development opportunities (Collaborative for Neighborhood Transformation, nd). The core principles of an ABCD approach include active individuals in the community, associations (such as neighborhood associations), institutions (such as the City of Orlando in this case), physical assets in the community, and connections between individuals and institutions (Collaborative for Neighborhood Transformation, nd). For the city, she said this approach:

> takes the stakeholders and then they look at the positive aspects of the neighborhood, and the philosophy is those stakeholders have the resources to resolve a lot of the issues that they have. What it is, it is changing your questions, changing your perspective rather than constantly complaining.

For example, she said neighbors often complain about traffic issues, so community associations can work with the city to find solutions. As it turns out, "most of the speeders are the residents so asset based community development would encourage you to use a little bit of peer pressure. You have the power to put that [slow down] sign in your yard."

A key tenet of this approach is sustainability or making the connections last beyond city involvement. In this respect, sometimes solutions to neighborhood problems might take a bit longer because the process relies heavily on the local stakeholders. "We know if the city would do everything for them, when we leave there's not the sustainability.

To me, a success is a sustainable neighborhood association." For the CNR team, developing strong associations is a vital part of maintaining and growing a strong overall city. Community leaders, she explained to me, are key because it takes someone to get the ball rolling and then a group to maintain that energy. "All of that takes time because you have to build trust," she said.

Since the neighborhood programming began with Mayor Frederick, there have been many changes as administrations shifted. In 1992, Mayor Glenda Hood consolidated many neighborhood services into one main office. The physical office space was intentionally not in City Hall at the time to better connect with local communities. A city employee in the Office of Community Affairs and Human Relations (CAHR), who has been with the city for more than 30 years, said that Mayor Hood's transition team had neighborhoods as a vital focus area. She said:

> As she described it to me, they wanted to do something that was empowering that would give citizens the tools that they needed to actually become more partners with government and to know how to access government services and programs but also how to use the various process and services that government had to offer to better their neighborhoods.

Before her election as mayor, Hood served as a city commissioner, and "she spent a lot of time visiting constituents in neighborhoods and communities," the CAHR employee told me.

> She also realized there were a lot of disconnects between what the citizens wanted to get done or needed to get done in their neighborhoods and what the city was aware of. The communication line wasn't open and clear as it should be.

One of her first jobs when joining the city in 1992 was to get a handle on what people considered their neighborhood area. Before that, the city relied on a rough idea of neighborhood boundaries. She said: "So when we started looked at developing the relationships with the neighborhoods, we looked at the geographic boundaries of the neighborhoods first and started to build on the preliminary map that the planners had put together to get more input from residents on what constitutes the beginning or the ending streets of certain neighborhoods."

From there, the office evolved as the city grew and the Mayor Buddy Dyer administration came into office. When Mayor Dyer was elected, there was a significant budget challenge in the city, so the neighborhood department was cut in 2003, with the exception of one person – the team leader. The CNR assistant director explained that in 2004, the city was able to hire two more staff members for the department, and today, there are three neighborhood outreach coordinators, the team leader, and a resource specialist (City of Orlando, 2018a). "What our team does is we let the neighborhoods know that we are representing the mayor and neighborhoods are very important to the city because if it wasn't we wouldn't have a neighborhood relations team," she explained. "His goal is to have every resident have a personal contact at City Hall, and he looks at the neighborhood relations team as being the group to make that happen."

How do neighborhoods create identity? Well, the assistant director not surprisingly said it depends. Take the Thornton Park neighborhood. Today, it is a thriving Main Street district, but it used to be run down. Only a handful of businesses were open there in the 1980s, including a 7/11 and hair salon. Then, some developers came in with an idea to turn the area around, buying older homes and either rehabbing them or tearing them down. "Thornton Park has developed over the years a reputation of being a trendy, kind of the happening place because it's walkable, it's just east of Lake Eola, and you have the amenities there." Interestingly, most of what people think of as Thornton Park is actually in the Lake Eola Heights historic neighborhood, so this shows that sometimes geographic boundaries matter less than what people think about as the neighborhood. "Because of marketing and they've always had an association, they have worked hard to try to increase their reputation."

Overall, the CNR team for the city acts as facilitators and liaisons to community leaders throughout Orlando with the goal of strengthening neighborhoods. They work with internal city departments to streamline communication materials for residents, ensuring any bureaucratic speak is lessened or, ideally, removed. Externally, the CNR team interfaces with leaders to build those sustainable associations mentioned earlier. "We are really active to make sure we are really responsive. I think there is intimacy, a sense of belonging. I would say that it's connectivity, like being connected to your city beyond what you can pull up on the web," said a member of the CNR team. Added another member: "Your home is probably your biggest investment, and you want the best community around you for that, and what does that look like for you? We help them define those goals to see how we can move forward, especially when they come together because they're so excited to do that."

Taking this philosophy a step further, the city launched iLead, a series of guides, webinars, videos, and other resources dedicated to helping residents build a stronger neighborhood (City of Orlando, 2018b). For example, some guides include meeting help (running meetings, parliamentary procedure, etc.), leadership skills, communications tools, avoiding burnout, and volunteer management (City of Orlando, 2018b). As one CNR team member said, iLead materials is "the stuff no one tells you" when it comes to being involved in your neighborhood.

Overall, the city believes in building strong neighborhoods through empowering community leaders to carry the torch. The CNR team is the "concierge" of the city, explained one team member. They connect policy and practice through actions and deeds. Said one team member:

> I love Orlando. We've got such great neighborhoods. You could live in Longwood, and if you are visiting someone you're going to say [you live in] Orlando. It's hard to find that ... Downtown definitely has the great restaurants and mom-and-pop and things that make it not a touristy area. You go there because you want to go to Lake Eola and experience everything.

This sums up the tension that is inherently Orlando – appealing to locals and tourists alike while trying to overcome the international image of inauthenticity given the theme parks. The neighborhoods have unique identities and brands that help bridge that gap. One way to do this is through Main Street districts that give neighborhoods a certain look and feel – an identity.

Main Street districts

Main Streets are a part of the city's overall economic development strategy. The Main Street program is run by Main Street America, an independent subsidiary of the National Trust, with a goal of preserving neighborhood character and boosting economic development (The National Main Street Center, 2018). As of this writing, Orlando has ten Main Street districts in various neighborhoods. City staff gladly work with Main Street districts, even having a dedicated staff member for helping district leaders. As one member of the CNR team said about Main Street leaders and volunteers, "They do it because they love their neighborhood, and they want to see more of the things they love going on. They volunteer in masses, they really do. People could be drawn to an area because of the district." I spoke with two Main

Street directors who explained why there is a brand identity associated with each area. (I should note that I reached out to all the Main Street directors for interviews, and only two scheduled with me.)

College Park Main Street district

The College Park Main Street district director was relatively new to his position when we spoke. He has a background in history and museum management and served as a volunteer on his local Main Street board in Cedar Falls, Iowa. As a director of a local museum, his role was to share with officials what was happening with community preservation efforts and historical landmarks in the city. "Developers will threaten older structures with the wrecking ball. I fought I don't know how many of those battles over the years, and I found Main Streets was always my comrades in arms over the year." He then went on to serve as Main Street director in Waterloo, Iowa, for five years before moving to Florida for his current position. "I was drawn to Main Streets originally because I love historical architecture. I love to see the unique, beautiful aspects of a community celebrated." Main Street districts, he said, sprang up as a means to preserve historic neighborhoods that were being abandoned as urban sprawl and malls came about, pushing out many mom-and-pop local businesses. In Waterloo, for instance, the Main Street program was set up to bring attention back to the downtown area when most people favored department stores and now even online shopping.

Given it is a national program, each Main Street district has certain rules and regulations to follow, but there is leeway in implementing new ideas. The City of Orlando has its own dedicated Main Streets liaison who helps leaders stay on track with goals and meeting those national requirements. As he explained it:

> Let's say a city wants to have a Main Street program, you have to be able to demonstrate that you have certain criteria in place. You have to have demonstrated support from the city government, from your downtown merchants, financial commitment to pay a program director. You basically have to have your ducks in a row as you define them.

College Park, he said, was developed in the 1920s with many single-family homes and quite car-centric. "I would make the case that College Park is almost like an individual little community nestled inside this big city, and they very much like to maintain their own identity." His district, and others, has four core committees: design, economic vitality, organization, and promotion. The design committee, naturally, focuses on the appearance of the district and community. A major long-term goal he has is to improve Edgewater Drive, the primary local business street in College Park. Currently, he said, sidewalks are too narrow with plantings in the middle. The street is set up as parallel parking spaces, a car lane, a turn lane, another car lane, then more parking spaces. "To cross the street as a pedestrian is a harrowing activity."

The economic vitality committee works with local College Park businesses to ensure they are thriving and open storefronts are filled. The district hosts at least eight networking events a year for the local business owners, and long term he is working on a business inventory of sorts to categorize what kinds of shops already exist and what could be added. The organization committee is responsible for the bylaws, finances, volunteers, and other bureaucratic responsibilities.

The promotions committee not only does the special events but also is working on an overall branding strategy for the district. He is encouraging team members to think what makes College Park unique compared to other Main Street districts that are a short drive away. He also thinks the logo could use some work. "It's easy for me as the new guy to be critical of it. I don't love it, and I don't think it really tells me anything unique about College Park. It could be anywhere in the U.S."

He said College Park is "fiercely unique" in that the district hosts annual events that draw thousands of people, including a jazz festival and Christmas celebration. That is what he likes most about his job as a Main Street director – finding and leveraging unique community assets. "We're all trying to connect to our neighborhoods, and I don't know how you connect if you go to McDonald's every day for your coffee. A place like this, it becomes part of your fabric as a resident." By "like this" he meant Downtown CREDO, the coffee shop where we met for our interview.

Currently, he said, the College Park brand is a small-town feel with single- and multigenerational homes. The brand, he said,

> is awfully hard to define in a marketing campaign … I can tell you what I thought as a new set of eyes. First thing that was sort of notable to me as I looked at is, why do they call it College Park? There's no college here. Then you see the naming of the streets. There's no real reason to name streets that way.

A challenge when trying to identify a strong brand identity, he said, is keeping that hometown feel as more multiunit buildings are springing up in the neighborhood.

As we met to talk, outside the windows of the coffee shop we could see construction of several apartment complexes, about four stories each. For College Park, those are high rises, so many locals are worried the aesthetic and composition of the neighborhood will change as more renters come in. For him, there is a tricky balance to strike because the apartments will mean more foot traffic into the local businesses, but "if suddenly there is another one of those things on the drawing board it's going to wipe out an admittedly 50-year-old suburban style."

As he thinks about the brand of College Park, he said that he wants to focus on being a more traditional neighborhood that is walkable and safe. "We are not going to be as artsy or funky or edgy as Mills 50. We're not quite Winter Park, but we're definitely not Mills 50 either."

Mills 50 Main Street district

Mills 50 is another Main Street district in Orlando at the intersection of Mills Avenue and Colonial Drive (also known as State Road 50). The area was once home to many Asian businesses and restaurants, and while some still remain, the area has a distinctly edgy feel. One article describes the area now called Mills 50 as one of the first major retail centers away from the downtown core (Davis, 2018). With the development of the Naval Training Center (what is now the Baldwin Park neighborhood and part of Audubon Park), the Mills 50 area became a popular shopping and dining neighborhood but quickly lost favor as Disney World came online in the 1970s (Davis, 2018). Vietnamese immigrants gave the neighborhood new life (Davis, 2018), and a rebranding effort was undertaken to focus on economic development through the Main Street program. In 2017, Thrillist, and online publication covering food, travel, and entertainment, named Mills 50 as one of America's up-and-coming neighborhoods, calling it a "cool" neighborhood that is "beautifully, blissfully tourist-free" (Meltzer, 2017, para. 4). Meltzer (2017) notes that the public art, eclectic local shops, and family-owned restaurants give the neighborhood its creative vibe.

The Mills 50 Main Street director said that rebranding the area took a while and was based on strategies that involved local businesses and residents alike. For her, she brought to the project a long experience with branding given her past positions with the City of Orlando running the original city arena and, more importantly for her, experience at Universal Orlando Resort. "I didn't know anything about branding until I went to Universal. That was a great education." In that position, she helped create a new logo for the popular Halloween Horror Nights event and aided in the City Walk entertainment district branding. She left Universal to start her own company and was soon tapped again by the city for her expertise. She joined the Mills 50 Main Street in 2010 after the then-executive director stepped down.

Her first task was to redo the old logo, which she said carried a lot of Asian feel that reflected the neighborhood's past rather than its growing future. She polled more than 30 local business owners to determine their thoughts on the neighborhood and how to work those into a brand identity, including a visual logo. She worked with a Mills 50 local business Orange Studio to create the logo that exists today. "It's clean. It's crisp. It's big. It's bold. People look at it and know exactly what it is." The logo is simply the words "Mills 50" in a clean font, usually on a black circle. "Mills 50 is much more than Asian restaurants on Colonial … When those [street] banners went up, businesses felt the area had an identity." Indeed, she noted that those restaurants are still a key draw for the area, and many have been around for decades. Other local businesses in the area, such as Track Shack and the Lamp Shade Fair, also have been open for decades, giving the area longevity as a neighborhood (Figure 2.1).

She describes Mills 50 as "hippy and funky" and it has a tagline of "intersection of creativity and culture." The tagline has been around before she began her position, but it still holds as authentic to the neighborhood because of the Asian influence and trendy local businesses. She called it a "foodie district" in Orlando, as the area is home to favorites such as (but not limited) to Hawker's Asian Street Fare, Pig Floyds (barbeque), and Tako Cheena (a late-night hangout). She spends most of her time working with local businesses that are the core of the Main Street district, ensuring they are happy. Given the area's popularity, a challenge is finding space for new business ventures and navigating the rules associated with Main Streets.

She also works closely with residents of Mills 50 neighborhoods including Lake Eola Heights, Park/Lake Highland, and Colonialtown North. She wanted to build pride in place and so encouraged people to use #Mills50Local on social media. Part of her idea was to engage local businesses to offer discounts and coupons for neighborhood residents. During one summer, she was able to get 45 local businesses to offer coupons or discounts, and as such got more than 100 residents

Figure 2.1 Track Shack.

to become members of the Mills 50 district, which normally supports businesses as members. Why get residents involved? She said that it is crucial for local business success, so promotions that actively seek those who call the area home bring people in during slower times to keep the area economically and socially vibrant. "It was a good marriage of the neighborhoods and businesses."

When I asked her why a neighborhood would want an identity, she spoke about community, property values, and businesses identity. "I think identity can be positive. They can also be detrimental." Part of the Mills 50 identity is its robust public art program. On its website, the district lists murals, Dumpster art, traffic boxes, and storm drains as its public art projects (Mills 50, 2018). The Dumpster art project is exactly how it sounds – artists adorn trash receptacles with beautiful paintings. The other public art projects are common across communities in Orlando and beyond, with art going on traffic boxes and storm drains. Usually, that art implores people not to dump junk down the drain. Events, too, make up a core part of the Mills 50 identity. An annual tour of homes has residents opening up historic homes for public tour, showing off the older architecture found in the neighborhoods.

There are after-hours networking events as well as community pub crawls (Mills 50, 2018). Events and art like these are important because "a neighborhood is more about the people in it than the actual boundaries or structures that are within it," she said.

Concluding remarks

Orlando takes a proactive approach when it comes to building its neighborhoods. It is not to say that other cities do not have similar tactics or departments. What I highlight here is that it is no accident that neighborhoods throughout the city have a certain identity – either created or earned. Some of those are positive (such as the Main Street districts), while others are negative (those areas that make the news for high crime). One of the hardest parts of the CNR team's job is wanting to help every resident with each complaint but knowing when and how to step in. There is an equity issue involved, as government cannot step in to solve every problem. The assistant director gave an example of when a new city commissioner heard of a woman in a neighborhood about to lose her house due to unpaid and unrepaired code enforcement violations. The commissioner wanted to step in to help, but there was nothing the city could do because the woman did not have insurance – and the equity issue was at the core. So, the commissioner activated church and nonprofit resources, and, as the CNR team member explained:

> got the house somewhat up to code, but it bit her because she had all of these people knocking on her door, and it was like well you did that for her. But it was truly her desire to help this woman, and it's hard. And that's probably one of the most difficult jobs we have as government employees to see that.

Yet setting priorities and connecting people is a big part of Orlando's neighborhood identity – from the city to the ground level. Talking to people in Orlando showed me the passion for place and the desire to create communities throughout the city so that people feel welcomed. Again, I am cautious to say this does not happen for everyone; people do feel alienated. But there are efforts and places to make a connections if someone wants such.

3 Emotional reasons for choosing a neighborhood

It looks like Desperate Housewives. I think when you first come in, it looks absolutely cheesy. When you become part of it you realize it's a really pleasant place to live.

– Baldwin Park resident

This chapter highlights some of the patterns found when analyzing interview data related to Orlando's neighborhood identities. Themes in this chapter fall into what I categorized as emotional reasons for choosing a neighborhood. A local Realtor told me that these emotional reasons are often more important than some pragmatic reasons when people choose where to live. This is in line with research that indicates how people perceive a neighborhood – the physical attributes – is important for feeling safe and trusting government institutions providing services (Greenberg, 1999).

The main subthemes in this chapter include sense of community, feelings of safety, and new urbanism. Each is explained through the voices of the interviewees. While I did not code it as its own theme, several interviewees underscored that Orlando is much more than its theme parks. One of the things visitors to Orlando find most surprising is how spread out the city is. People assume that if you live in Orlando, all the parks are a hop, skip, and jump away. Locals do not have to interact with many of the popular tourist spots if they do not want to, given Orlando's geographic distance. As one Baldwin Park neighborhood resident said about moving to Orlando from England, "Our expectations were quite low because when you say Orlando you think of theme parks. We didn't expect such nice communities."

Another Orlando resident, who was born and raised in the city, echoed much of these sentiments. She pointed out that at the Orlando International Airport, advertisements near baggage claim and throughout the airport try to point people toward other happenings in Orlando, such as outlet shopping or downtown Lake Eola events. Of course, though, people cannot escape the gift shops of Disney, SeaWorld, and Kennedy Space Center as they exit the airport. As the Orlando native explained:

> A year ago we had a conference at work, and it was running store owners from all over the country coming to Orlando. They were staying out near the Convention Center (close to the theme parks). I remember somebody saying, 'Why are we all the way out here? There's nothing to do here.' I remember saying 'This is not Orlando. This is tourist town.' If you really want to see Orlando, you need to come downtown or Mills 50 – really any area other than there.

One thing this resident also pointed out was that Disney World is not technically in Orlando, and she's right. There are many complex governing structures related to Disney's geography. If you look up where Disney is located, it is in Bay Lake and Lake Buena Vista, Florida. Reedy Creek Improvement District provides services to the area, and the District also has an interesting history. When Walt Disney selected the location for his theme parks, he selected the swamps of Central Florida because there was room to spread out. The problem was the land bordered two counties and was quite remote, with water and power lines about 15 miles away (Reedy Creek Improvement District, 2018).

To solve this problem, the state legislature in 1967 created Reedy Creek as a special taxing district to provide the same services as a typical local government agency (Reedy Creek Improvement District, 2018). This taxing structure limits the revenues each county collects from the Disney behemoth, and there is a constant stream of news articles about Disney's tax liability. In one case, the Orange County property appraiser assessed a Disney resort hotel, the Yacht and Beach Club, at $337 million versus the $189 million a judge ruled is the actual value after the company sued the county (Russon, 2018). As Russon notes (2018), these kinds of disputes are not unusual, and other local theme park operators SeaWorld Orlando and Universal Orlando Resort filed similar tax lawsuits against what they think are unfair assessment practices.

Another Orlando native, who now works for a developer in town, said that he never thought he would return to the city after leaving for college. But when he and his wife, who also is from Orlando, started a family, they moved back to Orlando to be closer to relatives. His father was involved in politics in Orlando in the 1970s, so, in his own words, "I'm more cynical" because he sees the intermingling of politics, development, and smart growth. He continued:

> It's a nice place to live because it's big enough that there are city things to do, but it's not all the hassles of living in a city you would have. It's a tradeoff. To me it's always kind of like, I don't know how to describe it, there's still some goofiness about Orlando. City officials want it to be so much bigger than what it is. It's always grow, grow, grow, grow. It's always wanted to be another city. It's not embrace [what is here]. It's, why can't we be more like another city? To me philosophically I don't quite understand it. It's really nice the way it is, and I want to keep growing it [wisely].

Even within city government, there is an effort to promote Orlando as more than a tourism destination. Explained a city official in the CNR office, the city started using #ThisIsOrlando "to differentiate that Orlando is more than Mickey Mouse." Highlighting Main Street business districts, neighborhood success stories, local art, and much more helped build this sense of identity away from only tourist attractions. Searching the hashtag on Twitter, for instance, shows people using it to tag photos of local restaurants, small businesses, news articles about Orlando, the airport, community events, and more. There is hardly a citing of roller coasters anywhere.

In these ways, Orlando locals try to create a sense of community and new identity apart from the theme parks. That is why neighborhoods are vital to that effort because each – even if they are not highlighted in this book – has a story to tell.

Sense of community

The first major theme that emerged in this section is a sense of community. Surely, this is not surprising, as people tend to build neighborhood attachment to a place where they feel connected. For those to whom I spoke, a sense of community was related to the place's identity. Integration and fulfillment are key aspects of why people remain in, and even advocate for, a community (McMillan & Chavis, 1986). A neighborhood sense of community has two critical elements: the physical place (neighborhood, street, city, region, etc.)

and relational elements (connection) (McMillan & Chavis, 1986). It is no surprise, then, that for the interviewees, a sense of community was critical to their feeling part of the neighborhood and the overall neighborhood identity. A sense of community includes membership in the community, influence within the community, need fulfillment, and a shared emotional connection (McMillan & Chavis, 1986). Many of the people interviewed for this book felt that in different ways. Others did not want that connection, and some people shared they will probably never feel that connection given their minority status.

I want to begin by sharing how two well-known Orlando neighborhoods – Lake Nona and Baldwin Park – created a sense of identity through planned design elements. Now, to be clear, there are others, but people from these neighborhoods generously spoke with me for this research. Baldwin Park is a former Navy training base, while Lake Nona sprang up as "medical city" for its proximity to hospital and other medical facilities. Both communities are quite different, yet they share similar demographics and characteristics. Residents in the two neighborhoods felt able to not only be part of the community but also help create it.

Lake Nona

I was able to get some backstory on Lake Nona to inform why there is a strong sense of community, especially in the Laureate Park neighborhood close to the retail core. A planner involved with the project told me "we had nothing else" when she started working for Tavistock, the development company that owns Lake Nona. Then the Department of Veterans Affairs (VA) hospital opened, followed by Nemours Children's Hospital. The rise of these hospitals led to others coming to town, thus the Medical City moniker. She took the project because:

> it was really creating a city from scratch pretty much. What I found intriguing about it is it's one of the only master planned communities that I've seen or read about that had more jobs than it had housing. That's one of the things that creates urban sprawl is when you just have houses and there's no place for people to work.

She explained to me that there is a relationship between place design and creating a sense of community. Laureate Park is the neighborhood closest to the downtown Lake Nona center, which includes restaurants, shops, a hotel, and businesses. There is another phase of development coming online, and in several years, Lake Nona will include a massive outdoor shopping space, movie theater, and more recreational activities. Currently, the U.S. Tennis Association makes its home out there, drawing more residents to the community for its sports and outdoor trail system.

But more than the physical amenities are the planned community events that help create a sense of community. "There are 1,000 events per year in Lake Nona," she said, including free yoga, food trucks, Tai Chi, local music, art in the park, and more. She noted:

> When you look at traditional neighborhood designed communities, sometimes it doesn't work out as well and it doesn't create the sense of community it's supposed to. Baldwin Park may be a good example of that because Baldwin Park doesn't have the strong programming that Laureate Park does.

Local restaurants, too, host events for residents including wine tastings and floral arranging.

Another feature that creates a sense of community? Mailboxes. Instead of each home having its own stand-alone mailbox, community planners grouped them together so that people could have another chance to interact with neighbors. "People get home from work generally at the same time, so that's another opportunity for them to meet their neighbors." In addition to mailboxes, porches are at eye level when someone walks by, putting pedestrians and homeowners on the same physical space. Interestingly, social media also have changed community design elements to foster a sense of community. For example, she explained that Lake Nona's community center has what she called an Instagram wall with a mural of wings. People can stop to take photos to share online. A stained-glass house serves as another hot photo spot in the community. These were purposefully integrated into the master plan to help create that sense of community (Figure 3.1).

The planner said that the identity of the Laureate Park neighborhood is "one of collaboration, and what I mean by that is people get together ... I think it's a very active neighborhood in terms of physical activity. There's always people cycling, there's always people walking the trail system." When I asked her why Laureate Park would want an identity, she explained that "it helps to level the playing field and gives everybody something in common."

Figure 3.1 Lake Nona art house.

The differentiation is key because neighborhood identity signals something to others about your place.

Baldwin Park

Another planned community is Baldwin Park, on the site of a former Naval Training Center, which operated from World War II until the 1990s (City of Orlando, 2018c). The training site was large, encompassing what today is Baldwin Park, the Audubon Park neighborhood, a shopping mall, and executive airport (City of Orlando, 2018c). When the base closed in 1993, the city began what it called the Base Reuse Plan to decide how best to use the land. After years of study and community involvement meetings, a final version of the community was underway in 1999 when city commissioners approved sale of the land to a private developer (City of Orlando, 2014). Baldwin Park would include residential units, restaurant and shopping space in a village center, recreation areas, schools, parks, office space, and more. Baldwin Park is divided into several smaller neighborhoods given the area's scale, and residential design guidelines dictate what kinds of housing can be built where (City of Orlando, 2014).

All the areas within Baldwin Park make for an interesting governing structure. The president (as of this writing) of the homeowner's association explained it this way:

> One of the challenges that was really sort of unknown was how do you govern a community like this? You have a residential owners association, and you have a commercial association. You have a CDD, which is a community development district, which governs common areas that are not part of the residential area like up Lake Baldwin Lane into Corrine [Drive, which is near the Audubon Park neighborhood]. The area by the lakes or areas around lakes are managed by CDD. Anything that touches water belongs to the city [like the] walking path around the lake and parks that touch the lake. The dog park is [neighboring city] Winter Park. Needless to say it creates some obstacles that often make it difficult to explain that you don't need to complain to the residential owners association about the commercial people blowing dirt off their sidewalks at 7:30 in the morning because the residential owners have nothing to do with it.

As noted, Baldwin Park is a planned community with construction and landscaping guidelines. For example, the residential land use guidelines are 163 pages, broken up by neighborhood area (ten total within Baldwin Park) and covering everything from garage placement to vegetation (City of Orlando, 2007). Said the association president:

> When they developed Baldwin Park, the primary interest on the part of the developer was to sell the houses in here, and the notion was we'll build this village center and when we fill the houses businesses will thrive. Well what they didn't think a lot about was some of the things that they did in creating that commercial association because they, too, have design guidelines and other requirements that they imposed on the community that have had a detrimental effect on commercial center development.
>
> So again, I love Baldwin Park. I just happen to know about all these warts that tend to affect the overall happiness of the clientele and the businesses here.

When I asked people why they lived in Baldwin Park, many of the answers were the same – family feeling, shops and restaurants, good schools, safety. Said one person who moved to the U.S. from England nearly six years ago:

It's clean, it's outdoors, it's friendly mostly. It's new, which reflects us coming to the USA, which was essentially a new start. So it's like a community that's got a new start. It's easy. Everything is easy to get to.

These are just two examples of communities that used their urban spaces to foster a sense of community with, some would argue, differing results. A sense of community was varied based on the person responding, with some wanting it, some not, and others feeling they cannot experience it (Figure 3.2).

Fostering a sense of community

Again, sense of community is a multifaceted concept that includes elements such as shared values, shared identity, and common interests in a small collective (Mannarini & Fedi, 2009). Sense of community is usually psychological and socially constructed (Brodsky & Marx, 2001). Sense of community does not have to be place specific, as people can find community in religious groups, school groups, book clubs, and more. A problem, though, is that community often

Figure 3.2 Lake Baldwin.

implies homogeneity and neglects diversity of people and views (Mannarini & Fedi, 2009). For the interviewees here, they expressed these varied senses of community within their neighborhoods. One Baldwin Park resident expressed sense of community as the following:

I can walk into somewhere and recognize people, and they'll recognize me. You can have a conversation. We may not be close friends or anything like that, but you just see these same people over and over again. When we walk our dog, there's other people who are out about the same time everyday walking their dog. At first it might be a wave, then a hello, then let's have a conversation about our dogs. You recognize the faces. You recognize the people, and you see them on a consistent basis. To me that's community.

Another person who lives in a more rural area of Orlando said that an ideal neighborhood (though not necessarily *his* ideal neighborhood) is one:

I think [with] similar interests, a sense of community. I like the idea of block parties. I like the idea of children being able to grow up together and being able to go to the same schools together and to be able to build those bonds.

It is no surprise that there is no one definition of sense of community. Here specifically, I look at research into a neighborhood sense of community as a means to pare down a vast literature. Sense of a community is a psychological concept and in this case means the attachment people have to their neighborhoods. Glynn (1981) was one of the first to attempt measuring this psychological construct that, he argues, had been taken for granted in earlier writings about sense of community. He wrote in a time when sense of community seemed to be on the decline thanks in part to industrialization and suburbanization. Explaining how he tested and developed the scale, Glynn (1981) found that a psychological sense of community can be both measured and manipulated. In other words, communities can do things to affect someone's belonging or exclusion. A positive sense of community in turn leads to more positive feelings of the neighborhood, thus also increasing its sense of identity.

This was often a recurring theme in this research. I spoke with a couple who raised their children in the College Park neighborhood and still live there today. College Park is aptly named because many of the streets are named after colleges or universities. (My colleague, for example, lives on Yale Street with her family.) This couple, with grown children out of the house, said that they picked College Park many years ago because "it is the feeling that there is a community, but in Orlando there's just so much. I guess things are so spread out that you don't feel it [the community]." For them, College Park was a way to get that sense of community within the neighborhood. The husband actually grew up in Orlando not too far from College Park, so he was familiar with the area when they were moving back to Orlando together to raise a family. They bought "the first College Park house that we felt like we could afford and then did a little renovation. And we're still doing renovation ... It was a nice street. I love the cobblestone streets."

Perkins and Long (2002) explain that neighborhood sense of community is an amalgamation of several aspects, including trust in one's neighbors, efficacy of organized collective action, formal participation in community events, and informal neighborly behavior. In other words, there is both a formal and informal aspect of sense of community, and each works together to build trust within the neighborhood and its organizations. These elements are logical as community usually means some kind of connection, though place is not necessary (think joining an advocacy group where you cannot attend meetings but still donate to feel connected). The same goes for local communities; they might not have one specific place but are geographically bound in some way (Chaskin, 1997). For example, one person told me that she goes to a certain farmers' market each weekend even though she does not live in that exact neighborhood. She goes because she feels a connection to that community rather than her block or geographic area.

Chaskin (1997), too, writes of the decline of community similar to Glynn (1981). Both are quick to say that decline can be real (economic depression that collapses business interests) or perceived (so-called broken windows, for example). Chaskin (1997) points out an exception: the ethnic neighborhood. Usually, these ethnic enclaves persist because people feel a sense of loyalty and belonging. But if neighborhood sense of community is on the decline, people seemingly are yearning for it in Orlando. As one resident of Baldwin Park said, a sense of community is:

> a connection with the place. It's the sense that you are part of it. I think it's the sense that you feel very safe and secure in that place. You're very comfortable. You're proud. You've got a very positive vibe about engagement. I like Baldwin Park. I feel part of it. I have a sense of loyalty to the shops and organizations even if they're not the best.

Neighborhood identity is a key part in feeling a sense of community. Vignoles et al. (2006) rely on the distinctiveness principle to explain why people form a sense of identity connected (or not) to place. Distinctiveness can be an individual or group project, meaning a person distinguishes themselves from one person or another, as well as one group or another. They argue that distinctiveness is a fundamental human need given we engage often in comparisons with one another. Related to place, one could identify themselves as a Londoner as opposed to a Berliner, a concept Twigger-Ross and Uzzell (1996) term place identification. Twigger-Ross and Uzzell (1996), building upon earlier work, indicate identity formation has four elements: distinctiveness, continuity, self-esteem, and self-efficacy.

Knez (2005) takes that theory and relates it to places, arguing place-related distinctiveness helps someone distinguish where they live (New York vs. Los Angeles), place-related continuity and congruency relates to why someone chooses a place (nostalgia, matches a personal perception), place-related self-esteem is someone living in a place that makes them feel good, and place-related self-efficacy means someone has what they need in a place. Based on his study of a Swedish city, he found that "persons considered to be highly attached to their residential area promote their place-related identity processes significantly more than those considered to be not attached to that area" (Knez, 2005, p. 214).

Much of this also manifested in my study. People felt connected to a place – a sense of community – if they were invested in some way. If people attended events, met neighbors, helped each other out, they felt that sense of attachment and identity. When asked, one person said that Baldwin Park "absolutely" matches her identity for this stage in her life because:

> From an age perspective, everyone on my block is within a year or two. There's probably eight families out of, let's say, the 13 on my street that are exactly our same demographics. That makes it nice for calling up and saying 'Hey we are going to go meet at this restaurant downtown.'

Identity, it should be noted, is fluid. It can change as you grow older and establish what you want from your neighborhood. Identity then can be used as a filtering process for choosing where to live – or where not to live.

To illustrate, for one resident of Baldwin Park, the sense of community still is a bit unnerving. Indeed, being a self-identified Latina woman of color, she expressed feeling unsafe at times when walking her baby in a stroller. She and her husband first moved to a neighborhood called

Hunters Creek when he was sent to work at Orlando Regional Medical Center. She chose the neighborhood because it was international, with a large Hispanic population, so she assumed that she would feel much more at ease. The house they rented, though, had many problems, including mold and an unresponsive landlord. She described that area as "just too busy, too high traffic" because it was close to Kissimmee and the theme parks. "That area was fine for the period of time we were at. We didn't have a baby at that time."

When she became pregnant, she knew she needed a change because she wanted good schools for her son. She and her husband have previously looked at Baldwin Park before choosing Hunters Creek but dismissed the area because "we were turned off by its demographic. It didn't seem as diverse to me. It's a very white neighborhood." Eventually, they decided to move to Baldwin Park and have gotten to know the community a bit better. As she explained:

> It's a lot more family oriented. I can go out and there's other mothers with babies. There's a lot of people walking an exercising by the lake. I like that aspect of it. It feels more like a small town rather than this huge dispersed space where you don't get any personalized feel.

While the community feels friendly to her, she explained that because she is Puerto Rican, and her husband is Pakistani, she still is hesitant to walk around alone. She said that one day, she was walking near what she called "really nice houses" and left her phone in the stroller, and was

> thinking 'God I hope nobody calls the cops on me.' As a person of color you worry about these things. I worry about that with my son. Because I had my son with me maybe I seemed a little less suspicious.

This fear is grounded in other negative experiences in Orlando, where people have told her to speak English when she was doing field research for her doctoral studies.

As she told me:

> It reminds you that this is still very much the South, and I think that's something that people forget about Orlando. Talking about identities, people don't associate it with the South; it's associated with the Mickey Mouse mentality and tourism, but people forget that. Racism is still very much alive, and there's still very much this kind of Confederate mentality. I mean we see Confederate

flags here. I don't think I could ever live here long term unless I did get a great job then I'll make sacrifices.

So, where did she grow up? The Bronx. What she described to me was exactly what Jane Jacobs (1961) detailed in her explanation of city life and why cities matter. Jacobs (1961) describes a scene where apartment dwellers or unit owners patrol the streets simply by looking out their windows. They recognize when someone is out of place. There is a sense that everyone is involved in looking after the block, the neighborhood. For this Orlando resident, her neighborhood in the Bronx was more ideal because "you would definitely see people you knew walking around the streets every day. People would definitely look out for each other." She left the Bronx to study at Villanova, and it was there she appreciated that people had different points of view – and came into contact with racism because she was from a different racial background and socioeconomic class than most of her peers in school. As a doctoral student, her interest is in political activism, so she attended the Women's March around Lake Eola in downtown Orlando. She said:

> People were so hesitant to call it a protest. I'm like, well what is it then? They didn't want to call it a protest. It was just a hot mess. I was like, why are we so confined to this space? In New York we wouldn't have cared about permitting issues. We would have taken it to the streets. I don't know if it's like organizing here feels different than organizing over there.

This person was quite candid when explaining why she moved to Baldwin Park from an area that was more Hispanic, more diverse. She told me she originally wanted that kind of neighborhood when she and her husband moved to Orlando but did not really know where to go. She asked people where there was a concentration of Puerto Ricans, so she moved to that area. When her son was on the way, she decided to move to an area with better schools and less traffic so that she could walk him in the stroller. In their old neighborhood, walkability was nonexistent. The Baldwin Park perception as a neighborhood for rich white people remains a constant sense of tension for her.

"I'm not talking about crime when I talk about Baldwin Park because there's a low crime rate here, and that's attractive," she said.

> But at the same time there's danger for me as a person of color in other ways. I've been in white spaces enough. I'm aware of it. I have training as an anthropologist so I can rationalize things.

I can be critical of these things, and I think for me having my son grow up in an area where I can take him out and we can walk and more family oriented and the schools and whatnot, I rationalize that in my head as, okay this is more important in this moment.

I share this story in length because it is important to highlight that for many, a sense of community is still elusive. This is lived experience that perhaps hits close to home for some readers or could be a surprise to others. She told me that she is active in Baldwin Park events, bringing her family to planned community happenings, such as the First Friday event featuring local vendors selling their goods. Even in those public spaces, she still negotiates feeling like she belongs and being wary of her surroundings. This is a lesson we can all take when explaining a sense of community, knowing it is not accessible to some in ways we might think.

Feelings of safety

The second theme I coded as an emotional reason for neighborhood attachment is feelings of safety. The key here is feelings, as many of the interviewees expressed safety as more of an emotion or sense rather than based on actual crime data for their neighborhood. Only two people told me that they based their answer on crime statistics from the Orlando Police Department. For everyone else, safety was how they felt walking around the neighborhood alone, or if they felt okay sending their children out to play.

It is perhaps not surprising that feelings of safety increase one's neighborhood attachment (Comstock et al., 2010). As with sense of community, there are formal and informal aspects of safety (hence the label here feelings of safety). Formal safety mechanisms include police patrols or neighborhood watch, while informal mechanisms can include self-monitoring or willingness to intervene in problematic situations (Sampson, Raudenbush & Earls, 1997). In their research, Sampson et al. (1997) found that social composition negatively affected instances of violence – the more perceived neighborhood control, the less likely there is for violent encounters.

As noted earlier, only two people cited crime statistics as the foundation for their feelings of safety. One was a woman who was searching for a new house with her family. They wanted an area with excellent schools where they also felt safe. She said that she and her husband made a list of must-haves and nice-to-haves, so found crime statistics for neighborhoods with houses they liked. The problem was that the housing market in Orlando at that time was so compressed that homes would fly off the market to other buyers before should could asses her list. So eventually, they abandoned the rational approach so they would not lose another house. They found a home in the Avalon Lakes neighborhood, a stark difference from their prior home in the Union Park neighborhood. For instance, she said her old house was along Colonial Drive, one of the city's busiest streets, and there was a homeless population living in the woods behind their neighborhood. As she told me:

> I think the cops ran them out of the woods so they wouldn't be back there, but now there are homeless people all along our streets. You have to lock your car at night because if you don't someone is going in if you leave your car unlocked every night.

Her new neighborhood is a gated community with a Walmart that is walkable. She also said she can walk her kids to school. While she had not yet met many neighbors when we spoke, she said that she and her husband had begun attending community meetings and hoped to meet more neighbors via the school. When she brought up safety, I asked her what that meant, and she said:

> I guess one thing is like, and I feel very like gentrifying saying this, but when I see people walking on the street if it's someone I can stop to talk to and not accost me. That's one thing I noted driving back and forth between the hoods. I left our new house at 10 p.m. and teens were walking to the ice cream place across the street. I was like, this would never happen in Union Park.

In their study, Delisi and Regoli (2000) asked how individual attachment to a neighborhood influences perceived safety. They hypothesized that people with fewer community ties, such as renters and loners, would perceive their neighborhood as less safe compared to homeowners or those actively involved in neighborhood relations. Their hypothesis was confirmed, but they also found that gender was not an important factor in perceptions of neighborhood safety, while a person's education, age, and race did matter. As they explain (2000, p. 186):

> Persons who tend not to socialize with neighbors, want to move out of the area, and renters rate their neighborhoods as unsafe. This perception is not mediated by fear of crime or criminal victimization. Thus, the importance of neighborhood phenomena to individuals is as salient as those individuals' sociological characteristics.

In other words, neighborhood ties matter when it comes to perceptions of safety. Neighbors in Baldwin Park experienced feelings of safety quite differently. One said:

> I feel safe, and I feel comfortable with my kids cycling around without supervision, which again is very European. I think Americans are paranoid about safety. In part it's legitimate, but for the most part I have a very safe vibe.

Said another who has lived in the neighborhood for four years: "It's well lit, it's well patrolled. Through the neighborhood app that we have, the neighborhood watch, there's just a lot of people paying attention to what's going on."

Another neighbor, who serves on the community development district board and is the other person who accesses crime data because of that role, said she would not describe Baldwin Park as safe. She attributes some of her perception to the nearby hospital where patients sometimes walk close to her house in hospital gowns. Another factor, she said, is that Baldwin Park is known for having expensive homes, thus making the neighborhood in general a target for crime. She said that she knows of an incident where some people came to Baldwin Park in a car "that looked like it belonged" but were there to steal items. She continued:

> As far as safety goes I would not say it's a safe hood. I see the police reports at the CDD (Community Development District), and you're like, holy crap this goes down every night. I don't go out after 10 (p.m.) ever because we are a target. Like if I leave my garage door open for five minutes things will be stolen ... We're too big of a target. There are people just trolling our alleyways.

Converse to the Delisi and Regoli (2000) study, Austin, Furr, and Spine (2002) found that gender did have an effect on perceptions of safety, with women feeling less safe than men. This makes sense, given women are often taught to fear others. As the resident of Avalon Lakes told me, she feels safe when she thinks she can walk around her neighborhood without anyone approaching her for something, and "my husband doesn't seem to worry about this at all" when he walks alone in the neighborhood. Austin et al. (2002) also found that homeowners also reported lower levels of perceived safety, likely because they are hyperaware of demographic changes that they see as a threat to their livelihoods. Housing quality, be that single-owners or apartments, also matters for perceptions of safety (Austin et al., 2002).

The broken windows theory (Kelling & Wilson, 1982) supports Austin et al.'s (2002) findings. Their famous theory uses the example of a broken window to illustrate the connection between neighborhood disorder and crime. The theory goes that if one window is broken and left unrepaired, the others will soon be broken because it seems socially acceptable. Kelling and Wilson (1982) note that this is just as likely in so-called nicer neighborhoods as it is in lower income ones. They argue that a broken window is a sign that people do not care and thus an invitation to break others. A stable neighborhood of families and people who care for homes should, ideally, stave off broken windows.

For some, feelings of safety were closely linked to their sense of community. Said a resident of the Celebration community near Orlando:

> To me a neighborhood does not have heavy police or security presence, but it feels safe and secure. Most important in the neighborhood is people and their friendliness with neighbors. There's friendliness with people who are not necessarily neighbors but are walking by.

New urbanism

The final theme coded as an emotional reason for choosing a neighborhood is linked to new urbanism. For many Orlando residents to whom I spoke, a connection to the place came from its walkability and access to recreation, shopping, and dining. To be clear, this was not everyone to whom I spoke. For some in Orlando, they wanted to escape a busy new urbanist lifestyle for neighborhoods that were more spread out and had land between houses. Though for the majority, a big part of their attachment to the neighborhood came from new urbanist features.

New urbanism typically refers to planned developments that include mixed housing options, walkability, and shopping that can all ideally be reached without a car. New urbanism traces its roots to developers Andres Duany and Elizabeth Plater-Zyberk, and they explained that their idea started as an aesthetic movement to create nicer neighborhoods but morphed into a public health movement related to climate change and personal health (Redmon, 2010). The idea was to shift back to more city-like thinking in design rather than develop neighborhoods and businesses that were car-centric – hence new urbanism. A problem, though, is when new urbanism runs rampant and people destroy what was built (Redmon, 2010). Ideally, new urbanism with its compact design should counter growth-for-growth's sake (Duany & Plater-Zyberk, 1992).

New urbanism is not without its critics, and this is not the space to delve into all of them. Ellis (2002) synthesizes the debate, noting three major concerns often include empirical performance of the communities (how well they actually achieve their goals), ideological and cultural issues (who can afford to live there), and aesthetic quality (could trend into cookie-cutter developments that look the same). Ellis seems to favor new urbanism, calling many of its portrayals in the literature caricatures. Related, Gordon and Richardson (1997) also outline some challenges associated with new urbanism, including the real return on investment of compact development, actual traffic congestion if corporations follow suburban developments, and technological innovations that might reduce the need for compact development (such as online shopping vs. shopping small businesses).

Despite some of the challenges, many people to whom I spoke appreciated a sense of new urbanism – or at least the ability to have a walkable neighborhood. Orlando is quite large and spread out. Anecdotally, visitors are often surprised I tell them I live about 40 minutes from Disney World. They say things like, "Oh I thought you lived in Orlando?" Well, I do, but the city is large, about 113 square miles. Finding walkability is a challenge, as Florida has high pedestrian fatality rates (Retting, 2018), and Orlando is constantly one of the most dangerous cities for pedestrians and cyclists (Shuler, 2018). Said one Baldwin Park resident, "The area was, and this is kind of like my gripe with Orlando in general, I just don't like how kind of spread out Orlando is. That you have to drive everywhere. I hate driving, especially I-4 gives me a lot of anxiety." (I-4 connects Orlando and Tampa and is one of the most dangerous roads in the country (LaGrone, 2017).)

New urbanism, though, is seen as a potential solution to the problem, advocating more density and less vehicular traffic. The merits, though, of these traffic-calming effects are indeed up for debate (Crane, 1996), but emotionally people seem to enjoy the idea of having a walkable space. A problem, though, is taking ideas of new urbanism and trying to apply a one-size-fits all approach to design (Cozens & Hiller, 2008). Ideally, new urbanism should take into account community dynamics and city dynamics to function well. For example, one Baldwin Park resident said:

> It's a great place to come hang out, have something to eat. There's not a lot of retail here. You're not coming here to just shop, but it's a great place to come and sit outside, have a glass of wine or coffee, and come hang out.

That is slowly changing, as Baldwin Park has a décor shop and clothing boutique. The problem, though, is that in line with new urbanism literature (Cozens & Hiller, 2008), the retail is not evident from the street so people would have to purposefully seek it out.

The Baldwin Park residential association president spoke about this concern, telling me about a class project students at Rollins College in Winter Park completed several years ago to study how to better market the neighborhood. He described the project as having five teams of five undergraduate marketing students develop a plan to promote the local Baldwin Park businesses. As he said, during the time of the study "the Village Center hadn't quite flesh itself out yet. It still hasn't for that matter, but it's a lot closer than what it was." He said that the students were able to find public records for the management company and calculated that each local Baldwin Park resident needed to shop at least three times a week in the Village Center and spend $170 per trip. "So needless to say that wasn't going to happen."

An idea was to broaden the marketing, then, beyond Baldwin Park residents, so one idea was to hinge on the lake. One student group suggested a marketing tagline of "lakefront shopping and dining" because, "they were pointing out there are not other places in Orlando that really feature that sort of lakeside shopping and dining." New urbanism features make it convenient for people to reach some of these shops and restaurants, but those who do not live in the neighborhood need to seek it out. Whereas, said the association president, nearby Audubon Park has a prominent shopping district on a main street called Corrine Drive. The businesses are more accessible that way. There's a tradeoff.

A local Realtor said that people expect more from a neighborhood today than they did in the past. For him, he sees a demographic split when it comes to homebuyers. He said his single, male clients usually look for things to do in a relatively close proximity and want easy access to transit options (such as living near a highway). He said that his single, female clients typically want a neighborhood where they can build connections and friendships. Families, naturally, want good school districts. He said communities are being built today that have more features than the typical suburban neighborhood. In his experience, developers are focusing more on landscape, for example, as a way to create communities in neighborhoods. Streets, too, are being curved to add interest and dimension rather than only straight up and down. Potential homeowners, he said, usually vacillate between what I have termed pragmatic and emotional reasons for choosing a neighborhood. When they come to him, "they're articulating the stuff, they're not articulating any of the emotional side of it, which when buyers first start thinking that's what they're [going to live]." Another local Realtor agreed. "Seven out of eight times they don't end up picking anything close to what they told me in their initial consultation," she said. Sometimes her clients, she explained, will have one set of criteria in mind and then go see a friend's house and completely change their minds to want features they admired in that home. She will try to tell her clients that even though the home has the physical features, perhaps the school district is not what they want. "We definitely reel things in because as an agent people sometimes have buyers' remorse, and I want to eliminate that. I don't want to go through that ever."

For on local planner, architecture – whether associated with new urbanism or not – is part of a neighborhood's brand and identity. While she works for Tavistock and Lake Nona, she lives in Audubon Park, which she described as having 1950s, ranch-style, mid-century modern homes. "That's part of the branding of Audubon Park. Some houses are being torn down and track-style housing has been put up in its place. It's a hot topic on Nextdoor ... Your architecture contributes to your brand." As mentioned earlier, the Mills 50 district does a tour of homes to show off its classic architecture and historical homes. The reliance on architecture is not a surprising element of place branding given it is so unique and can define an area (2008).

All told, people to whom I spoke seemed to like the idea of new urbanism's features – having shopping nearby, connectivity, and reduced reliance on a car. Again, this is not a blanket statement to say everyone wants that in a neighborhood. Some people might not mind driving everywhere or living in a quitter suburb. It seems, though, aspiring to have elements of new urbanism is important for a neighborhood identity.

Concluding remarks

These were the main emotional reasons coded in the data for why people chose a neighborhood. These are key components of creating and promoting a brand and identity. Sometimes, this is purposefully done to rebrand, as in the case of Mills 50; sometimes, it starts from the ground up, as in Lake Nona and Baldwin Park. It should be kept in mind, though, that these are individual preferences. For some, new urbanism is not anything they want. For others, they prefer land and acreage rather than a sense of community in the neighborhood.

To show that, I conclude this chapter with a story from a longtime Orlando resident who moved here as a child in the 1970s after his father was transferred to the city for work. They moved to the Dr. Phillips area, which back then was mostly orange groves. (Today, it is one of the wealthier parts of Orlando close to the theme parks.) The Dr. Phillips area is also close to Windermere, today one of the wealthiest towns in the region. When he moved here, Windermere "was just fish camps. There were small bungalows on the lake. It wasn't until the 1980s where people started buying up the bungalows and tearing them down to build mansions and stuff."

He went to high school in Winter Garden and had to be bused in because he said that there were no other nearby schools. As a child, his goal was eventually to move into the Rosemont neighborhood

> because that's kind of where you made it. That's where the young professionals lived. It had a golf course, a country club with a bar that you could meet afterwards. It was a little community, and I got my first place there. I rented an apt in Rosemont. It was really unique.

The same neighborhood today, he explained, is on the decline because of the rise of the nearby MetroWest community. He said:

> A lot of the homeowners moved out of Rosemont, and then a lot of folks when they moved in didn't bring the right style. A lot of people just moved out. The golf course went bankrupt, bought by an out-of-country developer. They ended up fencing the golf course.

When he moved into another condo, it showed him he really wanted more space. He said that when he was younger, he really wanted that sense of community and closeness, but as he got older, having more space and time to himself was more important. He said:

> You get to a point where you're spending your hard-earned money and just living in anxiety because you never know who was sitting on the hood of your car. It just got to be a point where you decide that you can go somewhere else. That's why I bought the acreage so I could roll across the front yard and no one will bother me.

He now lives on a larger plot of land outside of Orlando and commutes into the city, especially downtown, when he wants to be more social. "In 30 minutes I'm literally on a dirt road with my little world."

I share this story to show that things change – lifestyles change, desires change, needs change. Yet the themes in this chapter can be germane to any point in life. In the next chapter, I detail the more pragmatic reasons people choose a neighborhood and how those contribute to an overall identity.

4 Pragmatic reasons for choosing a neighborhood

> The people who come here, some of them like the architecture, a lot of them like the community feel. There's tons of activities every week. You sacrifice a big yard but get more conveniences – shopping, schools within walking distance. All the public schools are on site.
>
> – Lake Nona planner

In this chapter, I detail what I call the pragmatic reasons people choose a neighborhood. These reasons contribute to an overall neighborhood brand and identity because they are often key physical features that people use to determine if they want to live in a place and to see if the place says something about their desired lifestyle for that period in life. By pragmatic I mean the dictionary definition of handling or thinking about things in a reasonable, sensible way. In this chapter, three major patterns emerged again: walkability, bikeability, and recreation, schools, and green space. I explain each here, and readers will see a natural overlap between the pragmatic aspects of neighborhood identity and the emotional elements detailed in Chapter 3. For instance, if someone picked a neighborhood for its good schools, feelings of safety were often closely intertwined.

Walkability, bikeability, and recreation

Neighborhoods can contribute to a person's overall health and wellness, so it makes sense people would want to make a purchase in a place they see opportunities to reduce stress. Indeed, one study found that people living in disadvantaged communities are at a higher risk of heart disease compared to neighbors in more affluent communities (Diaz Roux et al., 2001).

One feature of a walkable neighborhood includes greenways. Greenways often are paved trail systems or open land that promotes outdoor recreation (Crompton, 2001). According to Crompton (2001), trails are not only physical spaces that promote activity but are also key elements of how people *perceive* the neighborhood. In other words, they are an important part of the neighborhood identity.

> Much depends on perceptions of who the users of the trails are likely to be. For example, if it is perceived that the trail may facilitate the movement of economically disadvantaged residents through a relatively affluent neighborhood, then the trail may be supported by the former, but resisted by some people in the latter group, who fear a decrease in their property value.
>
> (Crompton, 2011, p. 116)

He reports that residents perceived a nearby trail as having a positive effect on property values and as a key selling point of the home.

Walkability often includes other elements such as bikeability and access to goods and services. In other words, can I walk or bike to something I need or want? With bikeability, as with other pragmatic aspects of neighborhood identity, there is a closely linked perception issue. A social culture often is associated with biking (Smiley, Rushing & Scott, 2016), and so people perceive less access or appreciation of cycling if friends and family are not supportive, as well as feelings of safety and perceptions of crime in the neighborhood that might curtail cycling behaviors (Ma & Dill, 2017). Biking, even within a larger social culture, often is an individual choice influenced by overall perceptions of cycling reliability, safety, and enjoyability (Handy, Xing & Buehler, 2010).

One resident of Baldwin Park said, "on the weekend all I do is get on my bike." My apartment neighbor (as noted, I also live in Baldwin Park) posts almost nightly on his social media fees about biking around the lake. Said another woman in nearby Audubon Park, "If you want to walk or you want to ride your bike or if you want to take your kids out, you can walk these streets because it's a mature neighborhood. It's a nice neighborhood to walk in."

Interestingly, some of the comments made about bicycles had a close link to nostalgia for some interviewees. One former city employee who lives in a rural area outside of Orlando told me that he remains friends with kids with whom he went to school. "We always looked at who had the most bikes out front. Whoever had the most bikes is the house where it was going on." This idea of biking to school is linked with neighborhood nostalgia, as people remember that from their child-hoods. Bicycles become a way to connect with a place in a different way than in a car or even on a train, so ties into a person's sense of nostalgia because there is seemingly something pure about a bicycle (Gilley, 2014).

A CNR employee said that the city is really starting to embrace and create bike trails. According to Bike Orlando, there are approximately 255 miles of paved bike trails with a 30-mile radius of downtown Orlando (Bike Orlando, 2018). According to a local Realtor, people shopping for houses today are asking for not only more features in a house but also easy access to green space. He said:

> They're more cognizant of greenspace, walkability, even though our health numbers, if you look at weight gain and stuff like that in adult populations, the reality is it's not going down ... On the selection of the houses, people, especially in Florida, they want outdoor living space. They want open floor plans, they want light.

The city does place an emphasis on outdoor recreation activities for people of all ages in Orlando. The Families, Park, and Recreation department's website lists 19 neighborhood recreation centers meant for everyone from children to senior citizens (City of Orlando, 2018d). There is a skate park and tennis center, while Lake Nona is now home to the United States Tennis Association's training facility called the national campus. Said the executive at Tavistock, the Lake Nona developer, the tennis facility has events each day and people can book courts to come play. The local University of Central Florida tennis team also is based in the facility. He said, "300,000 people went through the facility last year."

In Orlando, there is an interesting connection between recreation and community economic development. The city is home to professional basketball and soccer teams and regularly hosts international sporting events in its Camping World Stadium. The stadium, though, was once in a state of disrepair. It sits in the middle of downtown Orlando and to the west of the stadium in a traditionally African-American section of the city that has a rich history yet is shrouded in images of crime and poverty (LIFT Orlando, 2018). In 2012, LIFT Orlando was founded because local leaders, spurred by the redevelopment of the stadium, wanted to tackle the pressing social issues facing the neighborhoods surrounding the stadium (LIFT Orlando, 2018). Through various community outreach efforts, the neighborhoods became more invested in their development and participation.

City officials from the CNR team worked with the LIFT Orlando board regarding community revitalization efforts in neighborhoods including Lake Sunset and Clearlake. "This is a beautiful example of letting the community decide what they want to call themselves. They've chosen to call this area West Lakes" as their brand identity, said the CNR assistant director. It is an interesting example of how recreation activities can lead to positive neighborhood transformation, as a key part of LIFT Orlando's success is increasing participation in Florida Citrus Sports summer programming in the neighborhood from 3 percent of local children to nearly 97 percent (LIFT Orlando, 2018).

Schools

Not surprisingly, a major reason people choose neighborhoods is schools. Most parents want neighborhoods zoned for top schools. Said one Baldwin Park resident:

> I think if you have kids the schools are probably about 80 percent of your decision, probably more actually. School is everything for a lot of people. Certainly here you can live further out and have a much nicer or bigger property for a lot less money, but then you haven't got the good schools.

Private schools are an option for some parents, but public schools are part of a neighborhood's identity – for better or worse. Schools are hotly debated and contested when parents feel they are not serving their needs (Hankins, 2007). Schools are one of the physical components that often define neighborhood boundaries (Campbell et al., 2009) and can thus largely influence decisions to live some place or not. Schools were by far one of the top reasons people in my study chose their neighborhood.

According to one branding expert who lives in Baldwin Park, there are two definitions of a neighborhood brand: hard and soft:

> A hard definition in U.S. is almost like a school zone – the school tends to define the neighborhood, the neighborhood tends to define the school. It's a very clear zone so to speak. In a soft sense, I think it's a place that has a strong sense of identity and community across all groups irrespective of age. I think with the neighborhood, if it's a strong neighborhood you'll have a very strong sense of place, a strong sense of belonging.

Schools influence crime rates, with different kinds of schools leading to the increase in certain kinds of criminal behavior (Willits, Broidy & Denman, 2013). Elementary schools generally do not increase crime at the block level, while high schools tend to lead toward assault, theft, and narcotics (Willits et al., 2013). Similarly, LaGrange (1999) found high schools in a neighborhood led to an increase in vandalism-related crimes. Schools can become places to build place attachment, especially if children are taught about their areas and neighborhoods (Brown, Perkins & Brown, 2003).

The British citizen now living in Baldwin Park said his primary concern when moving across the Atlantic was good schools. He asked friends living in Central Florida where to go, and most people suggested Winter Park, a city near Orlando. Specifically, they told him to find a neighborhood where his children could attend Brookshire Elementary, a public school in the area. He said he and his family rented within the school zone then went to try to buy property, but it was during the recession so really nothing was available. "Baldwin Park ironically was our Plan B. We didn't really want to come here, but in hindsight it was the best decision we ever made." Baldwin Park was Plan B because it was not in the same school zone, yet they moved, given other good schools in Baldwin Park. He continued:

> Even now we still talk about the Brookshire community because when we first moved here we moved a long, long way, and we were so happy to find a place where we settled so quickly, which we didn't expect because that's not what you think about Orlando.

As Hayes and Taylor (1996, p. 3) argue:

> A house is a collection of desirable characteristics such as shelter, comfort, and location. Therefore, the price that buyers are willing to pay for a house should be related to the prices they are willing to pay for its component characteristics.

In their study of Dallas neighborhoods, Hayes and Taylor (1996) find that people are willing to pay a price premium for homes in top school districts. The top indicator was school quality, including how well the schools perform relative to others. Less important are student body characteristics in their study. "Our analysis suggest that this premium for school quality can be among the most important determinants of housing prices" (ibid, p. 6).

Downes and Zabel (2002) also determined that schools matter for neighborhood choice but found almost opposite reasons from Hayes and Taylor (1996) in their longitudinal study of Chicago-area schools. Their results indicate that test scores remain important, but that buyers do indeed care about expenditures per student and the racial composition of the neighborhood and, thus, the school (Downes & Zabel, 2002). As school classifications based on overall performance become more pervasive, this trend of choosing neighborhoods with top-performing schools will no doubt continue (Figlio & Lucas, 2004).

For the parents in my study, schools were a great way to build a sense of community given they tend to get to know the parents of their children's friends. Parents also said they would not consider moving from their neighborhood if the children were in a good school.

Several parents said that after their children graduate high school, they might consider moving. Said one resident of Baldwin Park when asked if he considers moving, "No. I haven't got the energy. We wouldn't consider moving until the kids finish high school, so that's another six years yet so it's really not on the agenda. I like it here personally and professionally." He continued, saying, "kids change things. Oh my gosh, everything I do is for my kids. It just sort of bing one day and you see things completely differently."

One woman said that to her, the schools are more important than on overall neighborhood brand:

> Like, I live in Audubon Park because it has great schools. I don't live in Audubon Park because it has East End Market, and it's a progressive liberal hood. I like that about it, but that's not why I live there.

A local Realtor echoed this sentiment, explaining that "people with kids typically are concerned with safety, schools, traffic because they're worried about the kids, and they want to know, they want proximity to things. They want the kids to be able to play like their life." In other words, there is a bit of nostalgia associated with good neighborhood skills if there can be a sense of community related to those schools and connections.

With Lake Nona, the Tavistock planner told me that the neighborhood tends to attract more upper-class families because of the nearby Medical City and overall image. She said that people report moving to Laureate Park specifically because of the community feel, architecture, and events. She explained:

What drives people to buy, schools are always first and foremost in my mind as far as what's driving neighborhoods. Either that or they're affluent enough where it doesn't matter. Like, hey I'll spend the money ... Sometimes they want to live for the convenience and pay for private school.

Indeed, within Lake Nona, a school and YMCA facility was the cornerstone that drove demand. The Tavistock executive explained, "kids could go from school to after school programs at the Y. It really helped activate the neighborhood."

In the City of Orlando, the CNR team explained to me that they often work closely with the local school district, as the schools are integral to neighborhood success. One team member gave an example of the Colonialtown North neighborhood where the Orange County school system wanted to close Fern Creek Elementary School in 2014. According to a news report, two nearby elementary schools were set to open, and those schools would drop the Fern Creek population to about 200 students (WFTV, 2014). Parents and local community leaders met with school board representatives asking them to keep the school open (WFTV, 2014).

In 2017, Fern Creek was again on the list of schools set to close down because of initiatives to build schools that serve kindergarten through eighth grade rather than separate elementary and middle schools (Martin, 2017). Instead of closing completely, the Fern Creek campus turned into a gifted magnet school called Orlando Gifted Academy in August 2018 (Postal, 2017, 2018). Said one CNR team member:

There is going to be life there instead of a shuttered school. So when you start seeing closed buildings, it sends a message that something is going on in this neighborhood, and people start moving out. They want to get ahead of that wave.

Green space

The final pattern I coded as a pragmatic reason to choose a neighborhood is green space. Green space is closely linked with walkability and recreation regarding neighborhood identity. Oftentimes, people use the green space for recreation purposes. Neighborhoods that made green space an explicit part of the identity and marketing seemed popular among those interviewed. Green space can take many forms, including active parks, passive parks, and trail systems. For example,

Lake Nona markets heavily its 44 miles of trails. Baldwin Park relies on its lake and 2.5-mile recreation path.

It is perhaps not surprising that people sought green space and the ability to use that green space as open space is often linked to positive health outcomes (Sugiyama, 2012). Using Toronto as a case study, Kardan et al. (2015) found that people who lived in areas with high densities of trees reported a higher perception of health and physically showed fewer instances of heart disease. Their case study is interesting given people do not likely associate Toronto with sprawling green spaces. What they show is that even streets lined with trees increase peoples' perceptions of their own health and well-being. Not only are there health benefits, but access to green space also has been shown to reduce aggressive behavior in teenagers (Younan et al., 2016).

One woman said she chose Audubon Park because it was "rustic, green, and sustainable," pointing to the nearby parks and nearby Leu Gardens as key features of the neighborhood. She said that she also goes to Winter Park for the popular boat tour around the chain of lakes. People on the tour get to be outdoors and see the large mansions dotting the lakeside. One CNR team member also recognized the brand of Audubon Park as "gardening and sustainability." That contributes to the neighborhood identity because, as she said, "I think people will go there specifically so they can be with people who enjoy the same kind of thing."

City officials in the CNR department also hear many resident concerns about green space. One woman shared a story of a resident in Orlando calling to explain she was worried about newer development around her affecting trees. While the city cannot control much of that development, she said that they can "connect people and bring them into the same room. We can bring developers and the residents together to sit at the table to increase some greenspace."

Access to green space often is seen as a privilege. In their study, Powell et al. (2006) explain that neighborhoods with majority-minority residents were less likely to have access to green space and commercial fitness facilities. Similarly, Singh, Siahpush, and Kogan (2010) found that teens, especially females, living in poor socioeconomic neighborhoods with little to no access to safe outdoor spaces suffered from increase obesity compared to teens in more affluent neighborhoods.

In Baldwin Park, the central green space feature is the lake. Lake Baldwin has around it a 2.5-mile recreation path. It can easily be seen how green space and recreation go hand in hand, but people seemed to mention them separately, so I did the same in this writing. The Baldwin

Park homeowner's association director said that when he and his wife moved into Baldwin Park, it was just being built as a community. He explained:

> If you walk around the lake, when we first moved here it was probably only about maybe a third of the way built at that time. We would think to ourselves, 'Well I sure hope people walk on this, which is what it's supposed to be.

A personal aside here as Baldwin Park is my neighborhood. I lucked into an apartment that overlooks the lake. I had no idea what building or what view I had when I signed the lease, as I was living out of state at the time and could not be in Orlando to look at places to live. My parents did that for me, as they live about three hours south of Orlando. When a unit came open in my apartment complex, I signed the lease without even thinking as my parents quite liked the neighborhood, complex, and overall location of Baldwin Park. I was pleasantly surprised when I moved in and could look out my window to see the lake. There is always someone, as the leader hoped, using the path to run, bike, or walk. Seeing people out motivates me to put on sneakers and go run or walk. I cheered on a local 5K race from my balcony. There is something about seeing the lake that gives me peace.

In their study, Sugiyama et al. (2013) examined the connection between green space and health by studying neighborhood walking practices. They used a longitudinal study design to determine if green space is linked to maintaining a walking program through time. Using Australian neighborhoods, they found that while access to green space did not necessarily make the respondents initiate a walking routine, the green space did help regular walkers maintain their routines. They also found that participants in the study with access to green space were more likely to maintain a walking routine than those without access. Having access to a park helped neighborhood residents in the study maintain a walking routine (Sugiyama et al., 2013).

One Baldwin Park resident noted the connection between green space and sense of community. In Baldwin Park, there is a large recreation space called Blue Jacket Park where people gather to workout, play soccer and baseball, and use the basketball courts of the adjacent elementary school. Aside from this big park, there are smaller pocket parks and green spaces dotted throughout the community. He said that these hyper-local green spaces bring people together to create a sense of community and togetherness. "I think some is sort of environmentally initiated. You have nice walkways, pathways, sidewalks that lead to places that are close … You are never more than two blocks from a park in Baldwin Park." The green space at Blue Jacket also serves as a community meeting place for local events. For example, there is an annual hot air balloon festival that meets in Baldwin Park. There is live music and trivia nights. The park becomes a gathering place to meet neighbors and host others from Central Florida.

Similarly, a resident of Celebration, which is close to Orlando, told me that the parks were part of the reason he selected a home in the community. He said that when you come into Celebration, "you see the lush green, you see the well-kept parks." Celebration, like Baldwin Park and many other communities throughout the U.S. and globally, also uses its green spaces to foster a sense of community. There is a local farmers market and a popular marathon and half marathon run through Celebration. Parks are also built into the community design similar to Baldwin Park. Said one official with the Celebration government, there are more than 50 parks throughout the community, and "you are within half a mile of a neighborhood park in any house in Celebration." He also said that people come to Celebration for the schools, recreation opportunities, and nice community. "All these things are where people with like interests can get together as a community and really look out for each other as a community" (Figure 4.1).

Figure 4.1 Celebration.

What the residents quoted here mean is that it is not just that having a park matters; it is what is included in the park and green space. If green space is used to build community, it ideally has features that bring people together. Events are planned uses in the spaces, but on a daily basis, park features matter for bringing people out and together (Kaczynski, Potwarka & Sealens, 2008). Trails, either paved or unpaved, mattered for bringing people to use the parks, especially if those trails connected to a larger trail system (Kaczynski et al., 2008). Much of the research looks at the connection between green space and mental health. For example, in their study of green space and its link to stress reduction, Roe, Aspinall, and Ward Thompson (2017) found that people reported that going for a walk or seeking quiet helps them cope with stress. Green space also aided in positive mental health for stay-at-home parents, the elderly, and those in a lower socioeconomic bracket as the former two groups are more likely to take advantage of green space with potentially more flexible schedules, while the latter might seek green space to counter other potential lifestyle stressors (de Vries et al., 2003). In other words, developers and planners should see green space as a must-have rather an accessory when developing communities (Maas et al., 2006).

Concluding remarks

This chapter detailed pragmatic reasons people in this study chose a neighborhood and how those relate to an overall neighborhood identity. Each of the characteristics becomes something that can be marketed and communicated. Schools are arguably the most important reason on that list for parents, given they are so closely connected to neighborhood choice. Walkability, bikeability, and access to green space all also have positive effects on neighborhood attachment and branding. In Celebration, for example, their logo is silhouette of a girl riding a bicycle past a picket fence and trees with a dog chasing her playfully. The visual identity captures the essence of both pragmatic and emotional reasons for choosing a neighborhood.

Lake Nona, as mentioned, markets its extensive trail system, so health is a vital part of its brand identity. College Park presents itself as a more homegrown community, and the logo, albeit eclectic, shows that side of the neighborhood. The challenges associated with communicating a neighborhood brand identity relate to authenticity or keeping what is unique to the area. It is difficult when brand communications might not match reality, so there is caution when it comes to promoting these pragmatic elements.

5 #OrlandoUnited
Community cohesion after Pulse

I was speechless. These terrorist attacks are in other countries, but it was the first time we're feeling the same impact in your backyard. I was speechless. I saw friends who used to go to nightclub ... I needed to pack my stuff, clothing. I knew that I was going to do work for long hours, to provide support to the Mayor, the city, on this emergency situation.

– City staff member

June 12, 2016, changed Orlando forever. In the early hours of that Sunday morning, a lone gunman walked into the Pulse nightclub and murdered 49 innocent people. On that evening, Pulse, an LGBTQ nightclub, was hosting Latin night. The shooter declared his allegiance to ISIS, the Islamic State terrorist group, before engaging police in a long standoff and trapping people in the bathroom. When the evening was over, the 49 were killed and 68 were physically injured while simply trying to enjoy an evening of carefree fun.

It was a day that changed Orlando's sense of place forever. At the time, it was the most deadly mass shooting in the United States (only surpassed as of this writing by the massacre in Las Vegas that left 59 people dead and more than 500 injured). Cities globally came together to show support for Orlando, with social media playing a big part of that connection. This chapter highlights an overall community cohesion that emerged after Pulse. All of a sudden, any neighborhood distinctions – at least for the moment – became nonexistent. We were one community, and people who probably never thought they would connect came together to show support, courage, and love.

I debated with myself about including a chapter on Pulse in this book. It might seem like an aside to some readers, given the incident is not directly about neighborhood branding and identity. The Pulse tragedy, though, changed the way Orlando residents relate to each other – at least it appears that way. As this chapter will detail, the community came together in unity after the event, and there are still signs of victim remembrance all throughout Orlando. I included this chapter to show readers how identities can change quickly, as the tourism capital will now also be synonymous forever with one of the worst mass shootings in U.S. history.

The incident and city response

It is difficult to live in Orlando and not know someone who was personally affected by the Pulse tragedy. One of my friends lost her cousin. Many city employees lost friends that night.

Responding to such a large mass casualty incident took unprecedented levels of coordination, decision making, and empathy. For City of Orlando officials, this was uncharted territory. City officials train for hurricanes and other natural disasters, and had done drills for active shooters, but this was a totally different scale. An after-action report from the Federal Bureau of Investigation (FBI) found that Orlando Police officers followed protocols but that more training is needed to respond to terrorist attacks such as this (Lotan, 2017).

I was part of a University of Central Florida research team contracted to understand how city officials responded to the crisis. Our focus was not on police or fire but on everyday public administrators now turned into front-line responders. Some of the information in this chapter comes from that research, conducted during the first six months of 2017. The team consisted of myself, Dr. Thomas Bryer, and doctoral students Esteban Santis and Sofia Prysmakova. We conducted semi-structured interviews with almost 40 city and county officials who responded to the tragedy. Our liaison with the City of Orlando chose our interviewees, and we requested other interviewees based on suggestions during interviews.

One finding that really stood out was the neutral bureaucrat narrative went out the window. So often, public administrators are supposed to be value neutral to carry out their jobs. In the case of Pulse, this was extremely personal. Indeed, it was so hard for many to separate the personal from the professional that many interviewees expressed feelings of exhaustion during the initial response period. As one city official explained regarding their feelings:

> Personally and professionally, never dealing with something like this and why people are so evil. My husband has a phrase – you

can't understand crazy. You can't. We don't have that mindset ... Professionally, I mean the gravity of the situation was unlike any of us had dealt with before. We have homicides and murders and things but that doesn't require an (Emergency Operations Center) activation. Maybe an alert message text on weekend, or we hear about it on the news. It's not something we get involved with. Seeing what happened in the morning, in the news, it's like you've watched something like this in other cities and never thought about it happening here.

City of Orlando officials had many scenes of response – the nightclub itself, Emergency Operations Center (EOC), and in the days following a family reunification center, and a family assistance center. Long-term response efforts are still being carried out via a contract with Central Florida United Way. There is a recognition that metal health care will be needed in perpetuity for both victims and first responders given the gravity of what happened that night.

The City did a good job of making its public records available online, and they remain posted as of this writing.

Given that the focus of this book is on neighborhood branding and identity, the remainder of this chapter will highlight community cohesion efforts that have given Orlando an entirely different sense of identity and place. Looking around today, you still see the rainbow city logo in business windows or the #OrlandoUnited brand on cars, backpacks, laptops, and more.

Theme 1: this is personal

With Pulse, it felt personal. City officials relied on policies and procedures to respond yet recognized that employees needed flexibility and discretion given the fluidity of the situation. Within the EOC, this sense of community among city employees was evident when people simply showed up even if not directly charged with doing so. This was good because there were extra hands to help, but this also stifled things such as shift changes as the response continued. Said one city official:

Everything worked well that first day, taking pictures of my notes, sending back to the EOC. We were sharing information that way. As we set up in different locations, like the (Family Assistance Center), and had people in different locations, that's when communication was challenging.

For one city official, one of the fluent Spanish speakers on staff, Pulse was personal. He shared that he and his partner knew people murdered that night. For him, this no longer was only a job, but he had to represent both the LGBTQ and Latinx populations. He awoke that Sunday morning to a slew of text messages asking if he was okay, then he turned on the TV to see the news. "I was suspicious, nervous, anxious. They told me it was a shooting in Pulse nightclub. I was really worried because I have a lot of friends who used to go to this club."

His boss told him to report to the EOC given his communications function. For normal EOC activations, this person has a communications emergency services function (ESF). Given he speaks fluent Spanish, his role was to share information with local, national, and international Hispanic media. He also spent time translating city media alerts, press releases, and even Facebook posts. Given the population at the nightclub that night, Spanish speaking was a necessity not a luxury. "I got calls from all around the world. From Latin America countries, from Argentina, Chile, Ecuador, I don't know how they got my number."

He immediately went to work strategizing with those in the EOC about communication strategies, estimating he translated at least 30 documents into Spanish during the initial response phase, in addition to posting for the web and speaking with media representatives. Another big project with which he helped was recruiting Spanish-speaking volunteers. Admittedly, city staff did not have enough bilingual people, so volunteers were critical. He explained:

> I was the only one in the communication team in charge of media translation. Since the tragedy happened that morning, I was stuck in the operation center. I could not go out and be with the community and serve as the Hispanic spokesperson. It was critical for someone who speaks Spanish and talking to the Hispanic media to say what we're doing as a city government.

Many people who spoke English and Spanish bore an emotional burden, especially when translating death notifications at the family reunification center. They described hearing a death notice in English, translating it to Spanish, then translating again the family responses. He said of the translation:

> It was so fast, in a timely manner, so that stressed me out. On top of that, I lost two of my best friends that night, dealing with the work I was doing in the city, but at the same time feeling those

mixes of emotion because I lost two of my friends that morning so it was hard for me emotionally. Physically, I lost weight. I think that keeping my mind occupied, working on these issues, allowed me, a little bit, to be distracted from what happened to two of my loved ones who died.

Other people described shutting off emotions to handle the situation. Said one woman also helping with communication strategies:

I did not have anyone who went (to Pulse). My husband was next to me. I knew where he was. I don't know. I am pretty, yeah, I am pretty, like, I dig in, and I do the job, I can detach myself pretty well. I am probably still detached from it.

As soon as she got the phone call about the shooting, she jumped immediately into work mode, packing an overnight bag with pens, paper, a charger, and clean clothes with a city logo. Now she keeps a bag permanently packed in her home in case something like this happens again.

Another city official said that he received a phone call in the middle of the night but did not hear the phone ring. He woke up to countless messages about the Pulse shooting, ironically having only gotten home from an event downtown a few hours earlier. For him, the professional instincts kicked in quickly:

When that hearing that message there were several minutes when it was so real to me I couldn't do it. There was a point I was a zombie, just for those 5 minutes or so, but after it was okay you have a responsibility to just do and now that part of the public servant component kicks in. This is what we do ... My deal is to make sure the community is being served. That's the role, that's responsibility and kicking into that gear.

There are countless other stories like this, each illustrating the professionalism city officials had to use that day. It was a dark day, but the community cohesion was beyond what many could have imagined.

Theme 2: this was on purpose

The immediate communication response after Pulse was factual – what happened, why, how, when. City officials put out information about road closures. Trash pickup. The family reunification center.

The family assistance center. Death tolls. And more. This kind of information was vital to keep the community informed.

But then there was a switch. Mayor Dyer and his team made a conscious decision that hate will not define the community – a decision that would go on to shape the city's overall identity and brand regardless of neighborhood. The mayor's first press conference struck a tone of love, unity, and resilience. This was a bold step, given the scene went from Orlando Police to the FBI given the shooter's pledge of unity to the terrorist group ISIS. This was now a terrorist attack, prompting a large federal response and presence. One city official said that there was discussion and debate about what to say at the first press conference and even when to have it. "Collectively we decided to not have a press conference until the club and (shooter's) car were clear of explosives. We did not think it would install confidence to have a press conference and have a car blow up."

Mayor Dyer insisted that he be the first person to speak because "the community needed someone they know." In Orlando, there is a strong mayor form of government. The mayor has an active role in making and setting policy and is often the face of the city. Mayor Dyer is (as of this writing) the city's longest-serving mayor and a well-known figure in the community. As such, the decision was made for him – and his associated social media accounts – to be the face of the community during the initial response and recovery. Said one city official:

> We talked about what we wanted to do is give accurate, concise, factual information, restore calm, and reassure everybody that we had a handle on the situation and that everybody was safe and that's how we crafted the message ... Most communities often respond with fear, if something else is going to happen. Anger and hatred are also common emotions, and especially in this case it would have been easy to stereotype those of the Muslim faith in the community given the allegiance to ISIS. But Mayor Dyer knew we were better than that, that the community thrives on love and acceptance.

It was during the mayor's second press conference that he finally gave the official number of deceased – 50 including the shooter. "You could see it on the faces of seasoned journalists they were in shock." Before the number was read aloud, there was speculation that "only" 20 people had died. The mayor wanted to give factual information about the death count so the number did not have to change each time. Orlando Health, the local hospital where many of the victims were transported, said some information could not be given due to federal regulations. City officials needed to wait until official notifications were made before giving numbers to the media. "We were focused on the needs of the community, and the press was focused on learning about the killer and circumstances," said a city official.

A big part of the city's unity response was the #OrlandoUnited hashtag. In the wake of disasters, many cities will use #CityStrong. We saw this, for example, really take off after the Boston Marathon bombings. The #BostonStrong hashtag became a calling for strength and unity. After the mass shooting in Las Vegas, the #VegasStrong tag emerged. Indeed, the advertising agency responsible for the "What Happens in Vegas Stays in Vegas" campaign immediately took down those marketing materials for something more appropriate after the shooting at the MGM (Schultz, 2017). For instance, an ad with a black background thanked locals and visitors alike for supporting the City of Las Vegas (Schultz, 2017).

The importance of messaging after a tragedy cannot be underscored enough. Said one City of Orlando official:

> I would say that my primary goal was the repetitive message of not being defined by the act. I continued to focus people on how our community responded, and you can't create that on June 12th. We had been a community of diversity, equality, and inclusion long before that. Knowing I could call on the imam to deliver the right message, after 14 years you have the contacts Rolodex.

Orlando wanted to do a different route with its social media hashtag. Even though people did use #OrlandoStrong, the city's official tag was #OrlandoUnited. Along with the hashtag was a rainbow rendering of the city's logo, with the rainbow colors representing the LGBTQ community colors. A graphic designer for the city wanted to do something, so he created the design and sent it to his boss. His boss liked it enough to use it in official communications starting on June 14, just two days after the shooting. Said one city official working on communications:

> June 14th was very busy. It was a lot. Ordering lapel pens, stickers, buttons, with the logo just to hand out to somebody. We knew a big part of healing was people having something to show support. Having that was easy and everyone wanted it. People wanted something tangible to show support.

Figure 5.1 #OrlandoUnited logo.

The logo became so popular, such a visual for healing, that the city made it open access on its website. Anybody could download it to use how they saw fit. This was interesting from a branding perspective, but city officials said because the colors were not official Orlando tones, they were okay giving the logo away. Said one city employee: "The logo became such a huge thing. This helped people show their support, and people across the world were asking for it. It is something we still see to this day" (Figure 5.1).

Each year on the Pulse anniversary, there are continual messages of love, further adding to the city's image. As of this writing, the city has marked two anniversaries, and at each year, there has been something called Orlando United Day, whereby people are encouraged to do acts of kindness to memorialize the 49 victims. The county history center has a display each year of Pulse artifacts, as they saved and stored almost every item sent to the city to honor the victims. This includes cards, stuffed animals, candles, photos, and more left at the Dr. Phillips Center for the Performing Arts' (DPAC) lawn, which became the official memorial site in downtown Orlando. Around the globe, 49 bells often toll in unison as another reminder of what happened that day (One Orlando Alliance, 2018).

As one city official said of the community support:

> Everybody in the LGBTQ community was affected, whether they knew somebody or somebody's brother … But the response, the

people standing in line for 3 hours to give blood, the 10,000 people showing up for the first vigil at [Dr. Phillips Center], 40–50,000 people showing at up at Lake Eola park (was overwhelming).

Theme 3: the community showed up

One of the biggest logistical challenges that city officials faced was staging a family assistance center at the local Camping World Stadium downtown. The location was chosen because it was close to Pulse and pretty central within Orlando. There also happened to be no major events taking place in the days immediately following the Pulse tragedy. The center was staged two days after the shooting for a total of ten days.

There were many public and nonprofit organizations represented at the assistance center. For instance, airlines and the airport had a presence to help people get tickets to various countries for funerals for loved ones, as many victims' families lived out of country. A local pet charity was there for animals now left homeless in the tragedy's wake. Local mental health counselors also had tables set up. For the city official charged with organizing all of this in a matter of days, it was a logistical triumph. Security into the stadium remained tight to protect victims. Media were kept in another location. The Family Assistance Center was all about providing help quickly and humanely.

In addition to the official organizations on the assistance center list, many people from the Orlando community showed up to help. Unfortunately, many had to be turned away because of the security concerns, but the sentiment was recognized and appreciated. The same security concerns surfaced in the Family Reunification Center, which was set up at a local senior center in the hours following the Pulse shooting. Said one city official:

The FRC is where the FBI made the notification of next of kin. It was a very high-stress, very emotional facility. We had various chaplains for support groups, mental health counselors, volunteers to drop off water and food and to provide counseling support. The issue we had at the time, as much as people (wanted to help) we had to go with vetted individuals because you can't show up and tell me you're a counselor. I don't know who you are.

So, After Pulse, Orlando saw city employees come together, nonprofit organizations provide help, international businesses lend support, and the community outpouring surprised many. The first vigil began as a grassroots effort on Facebook. City officials got word of an event page for a vigil that attracted more than 25,000 positive responses of attending. Said a city employee:

> We can't have 25,000 people without law enforcement, EMS, water. It's June. At 7 o'clock at night it can be 95 degrees ... The community response was tremendous and overwhelming. [We] did not think about it and until it was happening.

The city waived the permitting fees for the vigil and stepped in to coordinate safety, logistics, and agendas. The EOC also needed to be activated to monitor the situation. Local traffic cameras were used to see crowd sizes and if even more resources were needed. One city official said that this level of response was:

> significant and overwhelming. We also had an insane amount of people and organizations that wanted to provide stuff. We had some organizations dropping off bottled water at fire stations. That was a little crazy at times. We are not as a city set up to take in donations. If somebody contacts us about wanting to provide donations, we redirected to Red Cross, United Way, Salvation Army. With this we had so much coming in we couldn't do that because it wasn't fair to Red Cross or Salvation Army. We set up a link on the city website you could click, type in, and send contact information. We could get back in touch with them.

Given the community response, city leaders chose the lawn of the DPAC as a temporary memorial site. It was open all hours, so people could come together to mourn, to leave notes, to reflect. The Orange County Regional History Center collected and preserved nearly all of the items left at the site and displays them during the shooting anniversary (Hudak, 2018).

One city leader underscored the importance of having public gathering spaces for the community to use after mass shooting events that are now all too common in the U.S., saying:

> This is how the community moved through this tragedy, coupled with these community-wide vigils and prayer services and

gatherings and public spaces. When I am doing my presentations [about Pulse], I point out the importance of having community spaces where people can come and collectively grief mourn and celebrate.

Theme 4: strangers to allies

For me, I woke up that morning in Toronto, ready to head to the airport for my flight home after a month as a visiting scholar in the city. It was a normal day, as far as I knew. I checked my Twitter feed and Orlando was trending for some reason. I quickly was able to find out what happened and began texting friends. Everyone I knew was okay. I made it to the airport and watched the news coverage unfold. I cried many times realizing that was my home on air. It was surreal. (I should add that I grew up in Coral Springs, Florida, which is right next door to Parkland, Florida. If both cities sound familiar, they were on the news a lot because that is where the mass shooting took place at Marjory Stoneman Douglas High School in 2018. I now have lived in two places home to two of the worst mass shootings in the country. To me, that is incomprehensible.)

When I landed in South Florida (where my parents live and my dad was picking me up from the airport), I asked my dad to immediately drive me to the local blood bank. Orlando needed donations. Unfortunately, I had not eaten enough that day to donate so went back the next morning and was able to give. I was not the only one. In Orlando, lines wrapped around buildings at many blood banks throughout the city. Volunteers even brought water to people standing in line for hours in the summer heat. Giving blood made people feel like they were doing something, anything to help. As one city official said:

> This became stronger than the sense of gloom and doom. You saw it the next day. You saw it with the blood bank, which was flooded and to give blood, bringing water to folks. It kind of moved out of the sense of shock or being paralyzed. It was 'We have to help.'

Orlando will be synonymous with Pulse. There is no escaping the reality. City officials try to pay it forward when they can, giving presentations throughout the world about their response efforts. The community outpouring is something they speak about often, especially how to handle the level of donations. Even for those who have lived and worked in the city for a while, the response was still unexpected. Said one official, "It was stronger than I thought. I feel like we had a good outpouring, but the showing beyond that really made me feel good about our community because it was deeper and stronger than I imagined."

Pulse was a small nightclub in downtown Orlando – a private space. Many of the mass shootings or mass casualty events in the U.S. are meant to send some kind of message, however warped. With Pulse, people turned into allies that night. Said one city employee who also is an Orlando native:

> I was pleasantly surprised to see that we didn't get caught up into distinctions between sexual orientation or race or ethnicity. Rather it was about these are people in our community. We as a community were impacted or hurt ... I did not see a pronounced or visible conversation about the things people may try to use to divide us.

Conclusion

I included this chapter because it seemed odd to talk about Orlando identity without mentioning what happened that early morning at Pulse. Orlando's identity as a city is so entwined with the event. Even as I write this in 2018, a new Pulse remembrance memorial mural debuted on the side of a local business. It depicts a girl blowing rainbow-colored, heart-shaped kisses. As a reporter described:

> The mural is intended to show the outpouring of love and support for victims of the Pulse nightclub shooting two years ago in Orlando, using the symbolic rainbow and hearts that have come to represent the tragedy at the gay entertainment venue. Pedestrians will be encouraged to stop and take a picture while they stand in the place of the shadow of a girl blowing the heart kisses.
>
> (Arnold, 2018, paras. 3–4)

The mural is one of many around Orlando. You cannot drive anywhere without seeing something connected to Pulse. Here on the University of Central Florida campus, there is a memorial painting remembering two students killed that night. Something that night shifted. At least it felt that way to me when I returned home from Toronto. I recognize that not everything is perfect – far from it. But we can try each day to come together a bit closer as a community to keep the positivity alive.

My friend lost her cousin that night. Many of us gathered to do a local 5K walk to raise money for the Pulse victims' fund. We walked with our friend to show support. We stopped at the nightclub, where a temporary memorial stands. The outside fencing has photos and quotes from the press coverage and community response. Visitors can look through the glass panels on the fence to see the Pulse building. You can read the names of many of the victims through the panes. The fence wraps around the building, still riddled with bullet holes from the horrors of that night. You cannot help but be moved in some way.

6 Conclusion

Theorizing neighborhood brands

People now, the younger groups go for the hot neighborhoods. Mills 50 is booming. You've got 10-10 Brewing opened. You've got the taquerias ... You have all these different businesses that are bringing people in because it is a good ZIP code, a good address.

– Former City of Orlando employee

The book until this point has showcased the results of a content analysis regarding what people think is important about their neighborhoods. Taken together, they form aspects of neighborhood identity, neighborhood image, and neighborhood brands. Some questions, though, still remain. First, how exactly does that neighborhood identity and image form? Second, why would a neighborhood need an identity? After examining those questions, I offer in this conclusion working definitions of both neighborhood identity and neighborhood brand based on a combination of these results and existing literature. Ideally, other scholars can take those definitions and refine them for future studies. Residents can use them to begin (or even counter) branding efforts in their neighborhoods.

How does neighborhood identity and image form?

To understand the answer to this question, I return to the brand identity concept introduced in Chapter 1. Brands are names or symbols that distinguish one thing (a place, product, company, etc.) from another. Corporate brands are usually easy to identify by associated logos, such as the famous Nike swoosh or Coca-Cola's signature red and white design. Brands become cognitive shortcuts in a sea of choices. Think about when you go purchase laundry detergent. The choices are endless, but many people likely stick with the same brand because it works, it is a value, their parents used it. Whatever the reason, the brand is a cognitive shortcut that allows you to zip down the aisle rather than standing there for hours, weighing the pros and cons of each detergent option.

Related, brand identity is the unique set of associations an organization tries to promote itself as having (Ghodeswar, 2008). As Phillips et al. (2014) explain, brand identity, too, is multifaceted and usually includes elements such as brand personality (like Apple is cutting edge and hip, and Volvo is safe) and brand equity (why people choose one brand over another). Typically, people associate brands with the visual images, and these slogans and logos are no doubt important (Phillips et al., 2014), but brands are also about creating emotional connections that go beyond the logo alone (Eshuis & Klijn, 2012). Brand identity is what the organization tries to put out about itself, while brand image is how users perceive that brand (Faircloth, Capella & Alford, 2001).

For this research, I asked people about both. I asked why a neighborhood needs an identity and to name some brand images associated with Orlando neighborhoods. For respondents, neighborhoods formed identity through design (new urbanism), reputation (negative or positive), and word-of-mouth (both pragmatic and emotional reasons). First, people cited the design as being important for neighborhood identity, and these manifested in the sections on walkability, recreation, new urbanism, and green space. Second, the reputation of neighborhoods formed organically (Audubon Park for some) or through strategic planning (Laureate Park for others). Reputation also was influenced by news coverage of neighborhoods, with many people readily identifying so-called good and bad neighborhoods. Finally, word-of-mouth (and these days word-of-mouse) helped build an identity. People tell friends to live in their neighborhood, and digital platforms like Nextdoor aid in building a reputation by what is posted.

First, neighborhood design is part of neighborhood identity by giving the place a certain look and feel. In the City of Orlando, as in other communities, signage programs are a major way to give neighborhoods their own identities. A city planner told me that his goal is to continue expanding the neighborhood signage program in the coming years. To get started, he sketched a rough map that geographically represents neighborhoods in close geographic proximity. His sketch does not exactly mirror the city's official neighborhoods map, but the point of his sketch was to get a larger view of the city's spread. According to his count, there are as of this writing 110 neighborhood identity signs in areas throughout Orlando. He keeps track of where each sign is placed with latitude, longitude, and notes (sign on median facing south, sign on pole facing north).

Another way city officials encourage neighborhood branding is through utility box art – or public art on traffic control utility boxes. Again, this is not unique to Orlando, but the art in each neighborhood reflects the overall brand identity. According to the city's website:

> Traffic art boxes allow neighborhood associations to brand themselves with a theme, encourage a sense of neighborhood pride, and brighten their landscape. Public art encourages a sense of ownership and pride within the neighborhood, and can be a small but important way of creating a sense of neighborhood identity.
>
> (City of Orlando, 2018e)

As highlighted in Chapter 2, Mills 50 is one neighborhood that has capitalized on these kinds of public art opportunities on utility boxes, storm drains, and Dumpsters to create a brand identity.

The reliance on design aspects related to creating neighborhood identity takes a concept we usually see in cities and places it in the microlevel. Think about the Eiffel Tower or the London Eye or the Sydney Opera House. These are all design elements that are unique to the place that cannot be found anywhere else (at least in the authentic form as Las Vegas and Epcot in Disney World have Eiffel Tower replicas, for example). The idea behind these sorts of mega designs is, ideally, to leave a positive legacy people remember (Evans, 2014). Ashworth (2009) explains the idea of a flagship building or design structure for a place is not new, mentioning the Coliseum in Rome and the Parthenon in Athens. Key to creating a brand identity via urban design are (1) having "notable and noticeable" architecture, and (2) using a relatively well-known designer especially related to signature buildings (Ashworth, 2009, p. 14).

Aside from a signature building, there is signature design (Ashworth, 2009), whereby there are design features throughout the area that give it a distinct look and feel.

> Signature design may be conveyed through an assortment of related buildings, spaces and streetscape elements, such as signage, paving, and street furniture which taken together make statements about the place. The objective, as with flagship structures, is not just a coherent unity in itself but differentiation and recognition.
>
> (Ashworth, 2009, p. 16)

For example, Baldwin Park, Laureate Park, and Celebration all look similar in design and feel, borrowing from the new urbanism mantra in the planning. The Baldwin Park residential building guide, for example, is clear on what features a home can and cannot have, and how many homes of similar look can be built near each other. Ashworth notes this kind of replication, though, can become problematic, as "the result is that the attempts at unique expression become replicated thus defeating the original objective" (p. 17; Figure 6.1).

Second, neighborhood identity includes, naturally, the reputation of the neighborhoods. A tricky question I asked people was how exactly this brand identity and image forms. Again, identity is what the neighborhood wants to project, while image is what others think of the neighborhood. Brand image is where people make their purchasing decisions, deciding on products or places because of a reputation. As Govers (2011) points out, there any many ways places can form reputations, including through leaders, partnerships, and popular media (now to include social media). It is a combination of factors that leads to the creation and brand identity and image.

I asked people to describe the brand of the neighborhoods, if there is one, and to name some other well-known Orlando neighborhoods

Figure 6.1 Lake Nona houses.

and why. This was a question to really understand the neighborhood branding and image-generation perceptions among locals. I asked a local branding expert to tell me some neighborhoods and what he thought about each (the brand image). I did not give a list but rather let him generate the names on his own. He said his brand image of Baldwin Park, his neighborhood, is "new, clean, accessible, inauthentic, spotlessly clean usually." College Park is "more authentic than here, older, stable," and Thornton Park is "quirky, central, arty, old, established, interesting." He described Laureate Park as "boring, beautiful homes, no atmosphere. I am not a fan as you can tell ... Baldwin Park has the benefit of being close to a diverse set of neighborhoods. Lake Nona is standalone so it's like Baldwin Park in isolation."

He said the Pine Hills neighborhood is "on the news a lot" because of "crime, crime, crime. I know people who grew up there and said it was a beautiful neighborhood." One longtime Orlando resident agreed, telling me "it was the crust. It was at the top of its day." He said that the neighborhood is where he went as a kid to watch movies or grocery shop with his parents (he was living in what is now the Dr. Phillips area close to Pine Hills). Pine Hills is where everybody wanted to be, he said:

> because Pine Hills was a predesigned development for Disney and Martin Marietta executives. That's why if you look at a lot of houses in Pine Hills they're two-story or split level. It was a planned area for that ... [Executives] came in in huge droves in the '70s and '80s, and they all needed places to live right away.

(Martin Marietta is a large building supply company with offices in Orlando.) As I was talking to this person, one of his employees came in the room, telling me he grew up in the Pine Hills area. "Our subdivision had probably 200 houses, and we were like the fourth black family there. My neighbors, they were my friends. There was no racism." Though he said that he and his friends went to a community pool and came home with a Confederate flag. Despite that, he said, "there were no problems" for him growing up there.

A resident of Thornton Park told me her open associations as well. She said:

> When I hear Dr. Phillips, I think of traffic. I think of sprawling and tourists. I would say on the flip side of that is Mills 50. I think of something that's up and coming, young people generally. A large variety. It's very versatile.

She said that the brand image of her neighborhood is "inviting, friendly," and College Park's is "I would say similar to Thornton Park. It has the neighborhood feel, walkable, safe." She brought up areas that have a more new urbanism design, saying those places could deter people from exploring other areas given they are usually insular. She added:

> I feel like a neighborhood like that, like Avalon Park or Lake Nona, is almost deterring people from wanting to get out or wanting to explore. I almost feel like it's a detriment to the people who do live there.

Indeed, some people who live in Baldwin Park often refer to it as the "Baldwin Park bubble" because almost everything you need is inside the neighborhood boundaries or just outside. Indeed, there is an Instagram account called @baldwinparkbubble that shows photos of people and places in the community (Instagram, 2018).

For Celebration, this identity question has been especially tricky given it is known as the town that Disney built. And this is quite true. The community began in 1994 and was developed by the Walt Disney Company. As such, Disney property is a short drive away, and if you walk or bike or drive through the neighborhoods, you will see many Mickey Mouse-shaped objects, such as lights or plants. Disney is in its blood. I knew I wanted to include Celebration in this book even though it is not in the City of Orlando because of this strong neighborhood identity. When I contacted someone from the city and they agreed to an interview, the first question he asked me was "Are you going to write anything negative? Because we don't give those kinds of interviews." I explained that I was a researcher simply trying to ask why the neighborhood has such a strong brand identity.

But a simple Google search told me why he was concerned. Symon (2017) wrote that "Celebration is so perfect, it's creepy" (para. 2) and quoted a resident who called it the Celebration bubble. Pilkington (2010) chronicles the housing downturn and a murder in his reporting of Celebration's downfall. He writes:

> Walking around the (center) of Celebration – they call it 'downtown' though it's not much more than a village square – is like stepping on to the set of The Truman Show, the film in which Jim Carrey plays a man trapped inside an invisible urban bubble.
>
> (para. 6)

There are other examples of this, and the homeowner's association president told me that he understands the potential negativity toward his neighborhood. He said:

"I think everyone knows this is the town that Disney built," he said. "Some would, I think if you polled them, some would say yea were happy with that and others would say, I wish people would think of us for something other than that. There might be a vocal small minority who would say let's see if we can find a new slogan or tagline or something." For him, Disney was exactly the reason he moved to the neighborhood. Indeed, he first heard about Celebration while on a Disney Cruise and invested in property even before retiring to Florida. He explained:

> This community will never, can never lose the fact that Disney did build the town, because that's a fact. We will never lose it. Could it be overtaken by something else? Sure ... Do I think Celebration is more than the town that Disney built? Absolutely. I don't think there's been a conscious effort to rebrand, but I think it could happen. I think a lot of that is organic. I think the process of choosing what the branding or identity might be.

Organic development is sometimes how neighborhood identities form. Said the planner who lives in Audubon Park,

> Once an area has a name people start identifying it by that name. I think the way the neighborhood is designed, I think is important as well. The type of businesses that are in a neighborhood I think help identify it a little bit. If it's got an amazing group of local restaurants, well that's part of its identity – we're going to this neighborhood because of its incredible restaurants.

She said Audubon Park "has a very granola and progressive and a ground-up identity" that formed over time and organically. Her quotes exemplify most of the patterns found about the relevance of neighborhood identity – that it is not one-dimensional.

The Main Street director in College Park spoke extensively about neighborhood identity. He explained that it usually takes five years for someone to feel acclimated to a place – to make friends, find a social group, find that sense of community, shop local businesses. That is "assuming the resident cares" to do this. Some people do indeed want to be left alone. He said, "people are looking for their tribe, this is where I fit in" so might choose a neighborhood that matches that personal identity. Connection is key, he explained, because in his experience that is what causes people to give up on neighborhoods personally and socially (not getting involved in events or neighborhood affairs, for example). He added:

I think people often live where they live because of two things. One would be their job. Or two because it's like their home, and they don't know what else to do and its terrifying for them to go somewhere else.

Neighborhood identity, though, is individual to a person, said the director. Each person is "wired so differently" that they want to live in a place that in some way reflects their stage in life. Schools reflect a certain stage, while perhaps access to bars and nightclubs reflects another. (Again, this is not meant to be a blanket statement or to suggest that only young people like nice bars and clubs. This is to mean that preferences can change.) While the neighborhood identity of College Park is still a work in progress (as explained in Chapter 2), the brand image might be more set in people's minds. "For College Park, I feel like we're kind of 'The Breakfast Club.' There's a little bit of everything. You can be any kind of person and find a home." He did note brands and images could have a negative side in that people could be excluded if they do not feel they fit or cannot afford a home there. (As an aside, Baldwin Park gets this reputation a lot given the mix of million-dollar mansions and single-bedroom apartments. The image of rich, white, and big wins out over the mixed housing options.)

Finally, technology, too, plays a role in developing brand identity today. Just as there is word-of-mouth, there is word of mouse, sometimes known as electronic word-of-mouth (Breazeale, 2009). Naturally, typical word-of-mouth communication is spoken between individuals and is fleeting (Breazeale, 2009), but electronic word-of-mouth usually relies on written text that can be more nuanced, thought out, and permanent (Sun et al., 2006). The key there is can be, given we all know to "never read the comments" on the internet because they are often mean. Nevertheless, online customer reviews and websites such as TripAdvisor (an online travel website) are becoming more important in today's brand landscape (Ayeh, Au & Law, 2013). Ayeh et al. (2013, p. 447, emphasis added) found that "online consumers are likely to employ user-generated reviews for their travel planning so long as they *perceive the source to be credible*—regardless of whether the review reflects the actual truth or not." In other words, these kinds of sites, including Nextdoor, are becoming more important when it comes to shaping a neighborhood identity.

I spoke with a representative from Nextdoor who described the platform as "the social network for neighborhoods." She said that they have a presence in all 50 states, with approximately 170,000 neighborhoods in the U.S. on the platform. The platform allows people to post recommendations, look for services, and ask neighbors for help and advice. "Nextdoor is very utilitarian, right? People join for health and safety," but the biggest use, she said, "is local recommendations." For instance, people in my neighborhood often post looking for local doctors, babysitters, pet sitters, and home services (plumbing, handymen/women, etc.). I used Nextdoor to find people willing to speak with me for this research. Public agencies can become partners with Nextdoor and use the platform to push information to residents. "A lot of people like myself, I mean Nextdoor is where I go first for any kind of thing that might be happening in my neighborhood. The first place that I [go is] Nextdoor because that's where the police department is talking to me, that's where the fire department is talking to me."

Tavistock, the Lake Nona developers, integrated technology into community planning, with the planning expert telling me "at Laureate Park, we have a gigabit community" because people want reliable, high-speed Internet connections. Tavistock even has its own telecommunications company that provides Internet, cable television, and home security services to residents. Social media, too, has changed neighborhood design, as the example earlier of Lake Nona's Instagram wall shows. The public art projects around Orlando also become sites for social media photos and sharing. She mentioned that planners now want to create "tweetable moments or these Instragrammable opportunities" in communities.

> As a planner we try to figure out where are those opportunities? How do u keep it classic so it's not just, it goes away? That's been an interesting change over the last ten years of how we plan spaces. You certainly don't want to have a space that looks empty or run down, and it's in the back of somebody's picture. It's like free advertising.

Within the City of Orlando's CNR team, one member said that technology "is an area our Neighborhood Relations team needs to be a little more aware of" because of the rapid developments. Social media technologies make it so that people expect an instant response to a question or complaint, and when an answer does not come quickly enough, "leaders feel like they're being ignored then call action reporters at TV stations to complain." Another team member said they try to ask neighborhood leaders, "Is that the image you want of your neighborhood? This is what all of Central Florida, a nine-county region, is going to hear about your Orlando neighborhood. So I say let's work together before you do that."

She said that the CNR team has seen a change in how people use social media platforms including Facebook, Twitter, and Nextdoor. To her point earlier, she said that people often take to these platforms to post photos of things they do not like in their neighborhood or post when car burglaries happen. These images, she likes to remind people, are permanent and build a negative image of the neighborhood for those seeing those images. To that end, "We encourage the neighborhood organizations to have social media accounts because of the younger generation, the 24/7 connectivity, the need for information, or not to have to call someone to get info."

The City of Orlando does have an agency presence on Nextdoor, and it seems they often post messages about city events. They also reply to comments on posts. For instance, the City posted about Thanksgiving Day residential recycling pickup being moved from Thursday to Saturday. A resident asked about the effects on Friday pickup services, and someone from the city (whose name appears on the post) replied that Friday services proceed as normal. Other posts include flu shot spots, hurricane season advice, the Mayor's City Academy, and more.

The reliance on today's technology to create a brand identity is not surprising, given technology has evolved to include mechanisms for two-way, interactive communication (Ketter & Avraham, 2012). The biggest change from a branding standpoint is the ability of users to help shape the brand, taking the power away from the organization and giving it to anyone with a smartphone or Internet connection (Ketter & Avraham, 2012). Despite the dialogic capabilities within the tools, place marketing and branding professionals might shy away from that interaction because of the noise involved (Sevin, 2013). Sevin (2013) focuses his analysis on Twitter specifically, finding that destinations use the tool to push information about the place while neglecting the interactivity. He reminds us, though, that Twitter has the power to reach many people in a short period of time, thus changing how places can brand themselves (Sevin, 2013). This can be positive given the cost of advertising or negative, as the CNR team officials said in their comments to neighborhood leaders posting about crime.

Why does a neighborhood need an identity?

Again, this might seem like a simple answer, but it is not. Identity and brands can be inclusive or exclusive, especially when it comes to neighborhoods where purchasing decisions are often involved. Someone might really like a neighborhood brand but cannot afford to live or shop there. Or someone might live somewhere they find a bit embarrassing so will claim another neighborhood. A local branding expert who lives in Baldwin Park explained:

> I think it reflects on you. It marks on you. You are where you live. So it's a reflection of your identity. There's an economic benefit, so for the most part property prices don't go up, they don't go down normally ... I'm not embarrassed about the neighborhood. Many people are. They disguise their neighborhood. They feel they need to trade up. I don't feel that way. Why? Because they feel inferior, they claim Winter Park rather than Orlando, for example. They feel a more positive association with the Winter Park crowd than Orlando.

I asked people what an identity can do for a neighborhood? The Lake Nona planner who lives in Audubon Park said:

> It helps to level the playing field, and it gives everybody something in common. You may not think that you have something in common with neighbors, but because of the identity of the community you do. It helps to differentiate you in the market so people want to live into Baldwin Park, Laureate Park, Audubon Park because there is an identity associated with those.

For Tavistock and Lake Nona, they take that identity quite seriously. Indeed, many areas surrounding Lake Nona have received cease and desist letters for trying to use the Lake Nona name and brand to attract people to the neighborhood. Lake Nona proper is 17-square-miles owned by Tavistock Development Company, a private developer (Lake Nona, 2018). According to Lake Nona's former chief marketing officer and current residential development executive, Lake Nona includes several neighborhoods (including Laureate Park), a country club, townhouses, and market-rate apartments. The town center includes restaurants and stores, with many more on the way. "When we're done it's going to be a city within a city in Orlando," he explained.

Medical City is really a core part of Lake Nona's identity, he said, because it has a high concentration of hospitals, including the University of Central Florida's medical school and a University of Florida-associated hospital. (Others were noted in Chapter 3.) The hospitals came to the site between 2008 and 2014 – the height of the economic downturn in 2010. Because Tavistock is a private company, it was able to weather some of that storm and continue building houses and attracting large corporations to the area. The first housing models opened in 2011, and they have been selling since then. "You brand all that, and it's an overused word, but we branded it as the next great place to live in Orlando with thoughtful design with the Orlando and Tavistock Good Housekeeping seal on it."

Lake Nona is a higher-end community, with the average home price about $500,000, the executive said. Residential surveys, he said, show that 90 percent of residents have a bachelor's degree, with 45 percent have advanced degrees. (This makes sense given the Medical City ties.) Many of the neighborhoods are family friendly, and the schools in the area are becoming better each year. Medical City is such an integral part of that identity, but it was not an automatic decision to lean into that branding. "You don't want to be known as Medical City. Nobody wants to live in Medical City but having it adjacent is a nice to have."

While the hospitals were coming online, there was a big push from then-Governor Jeb Bush for the state of Florida to get further into biomedical development, the executive explained to me. Tavistock tried to capture some biomedical companies, but that did not pan out for the long term. "We weren't sure if that was a message we could keep going on, Medical City." So, the company switched its marketing thinking to districts and overall health and wellness. Districts allow for marketing segmentation, the executive said. So, there can be the medical district, the sporting district, and the training district. That is how Lake Nona was able to attract the U.S. Tennis Association there, along with KPMG's world training headquarters. (KPMG is an international financial auditing company.)

When Lake Nona was first coming online though, marketing was a bit more difficult than it is today. In the early 2000s, he said, Lake Nona was known for its country club and people "knew rich golfers lived here." What began changing that image was a community called North Lake Park and the YMCA facility that opened in 1999. Then as the medical pieces started coming online, and more neighborhoods were under construction, the marketing changed. One campaign was called How Much Do You Nona while another was Lake Nona-ology, telling people there was so much more to Lake Nona than rich golfers. "It was a play on there is so much here and so much to learn about here it needs its own subject."

The marketing team did traditional approaches such as advertising, bus wraps, and billboards. "But as we grew it became more about proof points. Like okay, this is happening now," the executive said. Aside from showing and not telling, another hurdle the marketing team had to overcome was the perception that Lake Nona is so far away from central Orlando. Lake Nona is situated behind the airport, so people think it is far, but the community also is right off a major highway looping around Orlando. The highway is relatively new, so people were not sure how to get to Lake Nona before it opened. "For someone who has lived in Orlando for 10 years, they think it's out there, but people coming to the market don't think so." (For context, my neighborhood in Orlando is about 22 minutes from Lake Nona on the highway I described.)

Even for a community marketing professional, he struggles with the term and neighborhood identity. He really focuses, even in his current position, on placemaking – the integrated approach of community planning, design, and management. "What's the brand of Atlanta? I don't know. Orlando is tourism no matter what they do. We sell Orlando a lot," he said, noting that Tavistock has a team of employees who cold-call companies trying to entice them to move their locations to Orlando. "I do think a lot of the best success stories are fairly organic, but there is a whole industry of economic development that will tell you differently."

CNR team members walked me through how some Orlando neighborhoods formed an identity and associated image. One member talked about Edgewater Drive in College Park, a main shopping district with small local businesses. Edgewater Drive used to be a bit run-down and went through a revitalization in cooperation with the Main Street District there. Said one team member:

> Edgewater Drive today is beautiful, and I think they really played up their demographics … Edgewater Drive has that kind of more classic feel. You've got really nice furniture shops, restaurants. It's reflective of the folks who live there for sure.

Audubon Park, she said, formed its identity and green and natural after East End Market moved into the area. East End is a small, indoor market highlighting local vendors selling items such as cheese, cookies, smoothies, bread, coffee, and handmade leather goods and jewelry. "They've grown as a product of their demographic" similar to College Park, she said.

Neighborhood identities, though, can also be potentially damaging. Betancur and Smith (2016) use Chicago neighborhoods to show how what they call neighborhood repositioning is essentially a product of post-industrialization and economic shuffling. For them, branding and rebranding efforts are part of this strategic repositioning, so fixed geographic entities such as neighborhoods either benefit (or do not) from these branding efforts. This is the problem with gentrification efforts, they argue, that seek to replace one population with another, usually minority communities outed by wealthier, usually white occupants who have decided an area is now "hip and cool." "A precondition for gentrification is *managing the perceptions of a space*, usually presented in the public realm as slum and blight, in order to entice investors and dispel the fears among gentrifiers" (Betancur & Smith, 2016, p. 51).

This tension is from the outward and inward marketing of places and neighborhoods (Colomb & Kalandides, 2009). Using Berlin as their case study, they detail some of the positive effects associated with the "Be Berlin" marketing campaign aimed first at local Berliners to build pride in place and second at external consumers to come travel to Germany. The campaign was able to reimagine the way locals saw public spaces but also risks commodifying places and spaces once purely public, along with physically gentrifying neighborhoods people might now perceive as the place to be (Colomb & Kalandides, 2009). There is a danger in commodifying culture and making people fall into certain categories when it comes to neighborhood identity and gentrification (Hackworth & Rekers, 2005).

Keatinge and Martin (2016) examine this cultural appropriation and gentrification in their study of Toronto neighborhoods. They study a neighborhood called New Toronto in southwestern Toronto, focusing on the tension between gentrifiers who attempted to change the neighborhood and the emergence of a commercial district that did not meet those idealized conceptualizations. At the heart of the conflict was a "gentleman's club" called Mickey's. The authors explain the neighborhood was ripe of gentrification because of relatively inexpensive real estate and close proximity to the downtown core. While the neighborhood was changing, the economic district was further behind. They detail carefully the legal and political machinations residents took to alter the economic district's image, setting up a conflict between insiders and outsiders – those who embodied the neighborhood ideals versus those who sided with the strip club. The quote residents upset that their personal brand identity did not match the neighborhood brand image with the strip club present. The work highlights the inherent tension in branding efforts when it comes to issues of community and identity.

Working definitions

I draw on the two definitions of neighborhood branding and identity from Wherry (2011) and Masuda and Bookman (2018) to generate one based on the results from this Orlando investigation. I should be clear that this definition in no way supplants the others; rather it expands our theoretical foundations for this emerging area of inquiry and practice.

Neighborhood branding is an active, stakeholder-driven process that detects and promotes the neighborhood's greatest strengths for economic, social, political, and cultural capital while focusing on community cohesion. Embedded within the working definition are aspects of the emotional and pragmatic reasons for choosing a neighborhood. Outcome measures, then, should move beyond pure economic understandings of branding success.

Practically, this research adds to our understanding of how people are coming together to build social capital through branding and identity strategies. Neighborhoods are now wanting an identity in increasingly competitive markets and realize their collective power when it comes to remaining authentic (Masuda & Bookman, 2018). This book is not meant to be a how-to guide for those interested in neighborhood branding. Instead, it offers insight into how Orlando residents and city officials see neighborhood identity and brands. There is a complex relationship between a person and their neighborhood given the various factors involved in place attachment and neighborhood choice (Lewicka, 2011). Findings herein shed light on how people in Orlando perceive their neighborhoods, how they view other neighborhoods, and strategies the city is taking to create an overall healthy community.

The stories throughout show the complex interplay between governing entities and people when it comes to creating neighborhood identity. The CNR team said one of the biggest challenges they face, aside from finding engaged neighborhood leaders, is transforming government-speak into something accessible for all residents. Government bureaucracy can be hard to navigate for even the most seasoned experts. There are rules and regulations that government entities must follow, and if a resident is upset about a sidewalk or a neighbor's code enforcement violation, they might not realize the processes involved in correcting concerns. Internally, the CNR team, one member told me, is trying to work with other departments to streamline and ease communication with residents. "We are having people come from other departments to our office because it's such a great learning environment not only for how to communicate in a professional way but also with residents in a more outreach kind of way."

For example, CNR team members described public works projects in neighborhoods that were taking place without much communication, so residents were getting naturally upset. So, she said the CNR:

team was helping put together communication plans for new projects that were coming down the line. It got to be almost a full-time job because there's so many public works projects. They funded an outreach coordinator position who focuses specifically on public works.

Now instead of long letters or no communication at all, much of the relevant information on projects is put on a simple postcard that includes maps and pertinent project-related information.

The CNR assistant director told me her office is successful only because of residents and partnerships. Through various outreach and matching grant programs, the city works to strengthen all of its neighborhoods and form direct connections with people. "We don't see a lot of the same stuff residents see. Some residents have expectations the city should see that lifted sidewalk, but we're not looking for it. The neighbor who walks on it sees it every day." For many people, knowing how to report that sidewalk is a challenge, so the CNR team sees its role as being the go-to people for neighborhood outreach.

Concluding remarks

Throughout this book, I have tried to tell the tale of Orlando's neighborhoods through the lens of identity, image, and branding. This book begins to fill a gap in the literature regarding neighborhood branding studies, given this is an emerging area of inquiry (Keatinge & Martin, 2016). More studies can be done using other cities and neighborhoods as case studies to determine the dynamics in place there.

If there is one thing locals and natives wanted to convey it is that Orlando is more than only Disney World and tourism. It is a vibrant, growing city with its strengths and challenges like anywhere else. Said one member of the CNR team, part of her job is "trying to create that small-town feeling in a big city. We're not just Disney." Said an Orlando native:

> At the airport, when Buddy Dyer is talking, it's not just Disney and theme parks. He's talking about the [Dr. Phillips Center for the Performing Arts] and the technology that's being brought here.

She is referring to Mayor Dyer's recording on the airport shuttles welcoming people to Orlando or wishing them a safe trip hope, depending upon the tram's direction (Figure 6.2).

Figure 6.2 UCF's reflecting pond.

Overall, this book was an interesting project that helped me learn more about my new hometown. As a Florida kid, I will continue to enjoy my trips to Disney World but also appreciate the eclectic, distinct neighborhoods in the city that make up the other half or Orlando.

References

Arnold, K. (2018). New Pulse memorial coming to Mills50. Retrieved from: www.orlandosentinel.com/business/consumer/os-bz-lamp-shade-fair-mural-20180709-story.html

Ashworth, G.J. (2009). The instruments of place branding: How is it done? *European Spatial Research and Policy*, 16(1), 9–22.

Austin, D.M., Furr, L.A. & Spine, M. (2002). The effects of neighborhood conditions on perceptions of safety. *Journal of Criminal Justice*, 30(5), 417–427.

Ayeh, J.K., Au, N. & Law, R. (2013). "Do We Believe in TripAdvisor?" examining credibility perceptions and online travelers' attitude toward using user-generated content. *Journal of Travel Research*, 52(4), 437–452.

Bacon, E. (1975). *Orlando: A centennial history*. Chululota, FL: The Mickler House.

Barwick, C. (2014). Upwards, outwards, backwards? Residential choice and neighborhood use of middle-class Turkish-Germans. Working Paper, *Cities are Back in Town*, 1–20.

Bastias-Perez, P. & Var, T. (1995). Perceived impacts of tourism by residents. *Annals of Tourism Research*, 22, 208–210.

Baum-Snow, N. (2007). Did highways cause suburbanization? *The Quarterly Journal of Economics*, 122(2), 775–805.

Bayor, R.H. (1988). Roads to segregation: Atlanta in the twentieth century. *Journal of Urban History*, 15(1), 3–21.

Berg, B.L. (2001). *Qualitative research methods for the social sciences*. Ann Arbor: University of Michigan Press.

Betancur, J.J. & Smith, J.L. (2016). *Claiming neighborhood: New ways of understanding urban change*. Chicago, IL: University of Chicago Press.

Bike Orlando (2018). About us. Retrieved from: www.bikeorlando.net/about-bikeorlando.htm

Bolan, M. (1997). The mobility experience and neighborhood attachment. *Demography*, 34(2), 225–237.

Breazeale, M. (2009). Word-of-mouse: An assessment of electronic word-of-mouth research. *International Journal of Market Research*, 51(3), 1–19.

Brodsky, A.E. & Marx, C.M. (2001). Layers of identity: Multiple psychological senses of community within a community setting. *Journal of Community Psychology*, 29(2), 161–178.

Brown, B. et al. (2003). Place attachment in a revitalizing neighborhood: Individual and block level of analysis. *Journal of Environmental Psychology*, 23, 259–271.

Campbell, E. et al. (2009). Subjective constructions of neighborhood boundaries: Lessons from a qualitative study of four neighborhoods. *Journal of Urban Affairs*, 31(4), 461–490.

Candelaria, M. (2017). Conway chain of lakes binds old Orlando neighborhoods. Retrieved from: www.orlandosentinel.com/features/neighborhood-guide/os-conway-belle-isle-azalea-park-dover-shores-neighborhood-guide-20170625-htmlstory.html

Carpiano, R.M. (2008). Actual or potential neighborhood resources for health. In: Kawachi I., Subramanian S., Kim D. (eds) *Social capital and health*. New York: Springer, pp. 83–93.

Chaskin, R.J. (1997). Perspectives on neighborhood and community: A review of literature. *Social Science Review*, 71(4), 521–547.

Chaskin, R.J. (1998). Neighborhood as a unit of planning and action: A heuristic approach. *Journal of Planning Literature*, 13(1), 11–30.

City of Orlando (2007). Baldwin Park: A traditional Orlando neighborhood. Retrieved from: www.cityoforlando.net/Archive2016/BaldwinPark-ResidentialGuidelines.pdf

City of Orlando (2014). Baldwin Park/NTC main base: A brief history. Retrieved from: www.cityoforlando.net/city-planning/wp-content/uploads/sites/27/2014/05/BaldwinPark-History.pdf

City of Orlando (2018a). Office of communication and neighborhood relations. Retrieved from: www.cityoforlando.net/ocnr/who-we-are/

City of Orlando (2018b). Guides. Retrieved from: www.cityoforlando.net/ocnr/ilead/guides/

City of Orlando (2018c). Baldwin Park. Retrieved from: www.cityoforlando.net/city-planning/baldwin-park/

City of Orlando (2018d). Neighborhood centers. Retrieved from: www.cityoforlando.net/recreation/community-centers/

City of Orlando (2018e). Art outside the box – traffic box art for neighborhood associations.

Collaborative for Neighborhood Transformation (n.d.). What is asset based community development? Retrieved from: https://resources.depaul.edu/abcd-institute/resources/Documents/WhatisAssetBasedCommunity-Development.pdf

Colomb, C. & Kalandides, A. (2009). The "be Berlin" campaign: Old wine in new bottles or innovative form of participatory place branding? In Ashworth G. & Kavaratzis M. (eds) *Place Brand Management*. Cheltenham: Elgar, pp. 173–190.

Comstock, N. et al. (2010). Neighborhood attachment and its correlates: Exploring neighborhood conditions, collective efficacy, and gardening. *Journal of Environmental Psychology*, 30, 435–442.

Coulton, C.J. et al. (2001). Mapping residents' perceptions of neighborhood boundaries: A methodological note. *American Journal of Community Psychology*, 29(2), 371–383.

Cozens, P. & Hiller, D. (2008). The shape of things to come: New urbanism, the grid and cul-de-sac. *International Planning Studies*, 13(1), 51–73.

Crane, R. (1996). On form versus function: Will the new urbanism reduce traffic or increase it? *Journal of Planning Education and Research*, 15, 117–126.

Crompton. J.L. (2001). Perceptions of how the presence of greenway trails affects the value of proximate properties. *Journal of Park and Recreation Administration*, 19(3), 114–132.

Davis, E. (2018). An up & coming neighborhood: Orlando's Mills 50. Retrieved from: www.moderncities.com/article/2018-mar-an-up-coming-neighborhood-orlandos-mills-50

de Chernatony, L. (1999). Brand management through narrowing the gap between brand identity and brand reputation. *Journal of Marketing Management*, 15(1–3), 157–179,

Deener. A. (2007). Commerce as the structure and symbol of neighborhood life: Reshaping the meaning of community in Venice, California. *City & Community Development*, 6(4), 291–314.

Delisi, M. & Regoli, B. (2000). The individual and perceptions of neighborhood safety. *American Journal of Criminal Justice*, 24(2), 182–187.

de Vries, S. et al. (2003). Natural environments – healthy environments? An exploratory analysis of the relationship between greenspace and health. *Environment and Planning A*, 35, 1717–1731.

Diaz Roux, A. et al. (2001). Neighborhood residence and incidence of coronary heart disease. *New England Journal of Medicine*, 345(2), 99–106.

Dilulio, J.J. (1996). Help wanted: Economists, crime, and public policy. *Journal of Economic Perspectives*, 10(1), 3–24.

Duany, A. & Plater-Zyberk, E. (1992). The second coming of the American small town. *Wilson Quarterly*, Winter, 19–48.

Ellis, C. (2002). New urbanism: Critiques and rebuttals. *Journal of Urban Design*, 7(3), 261–291.

Eshuis, J., Klijn E. & Braun, E. (2014). Place marketing and citizen participation: branding as strategy to address the emotional dimension of policy making? *International Review of Administrative Sciences*, 80(1), 151–171.

Eshuis, J. & Klijn, E. (2012). *Branding in government and public management*. New York: Routledge.

Evans, G. (2014). Designing legacy and the legacy of design: London 2012 and the Regeneration Games. *Architectural Research Quarterly*, 18(4), 353–366.

Fallon, P. & Schofield, P. (2004). First-timer versus repeat visitors satisfaction: The case of Orlando, Florida. *Tourism Analysis*, 8(2–4), 205–210.

Faircloth, J.B., Capella, L.M. & Alford, B.L. (2001) The effect of brand attitude and brand image on brand equity. *Journal of Marketing Theory and Practice*, 9(3), 61–75.

Farris, J. & Kendrick, A. (2010). *Neighborhood brands*. Dallas, TX: BubbleLife Media.

Figlio, D.N. & Lucas, M.E. (2004). What's in a grade? School report cards and the housing market. *The American Economic Review*, 94(3), 591–604.

Ghodeswar, B.M. (2008). Building brand identity in competitive markets: A conceptual model. *Journal of Product & Brand Management*, 17(1), 4–12.

Gilley, B.J. (2014). Cycling nostalgia: Authenticity, tourism, and social critique in Tuscany. *Sport in History*, 34(2), 340–357.

Giorgio, P. (2017). Finding peace in Parramore: Neighborhood association president is making a difference. Retrieved from: www.clickorlando.com/getting-results-/award-winners/finding-peace-in-parramore-neighborhood-association-president-is-making-a-difference

Glynn, T.J. (1981). Psychological sense of community: Measurement and application. *Human Relations*, 34(7), 789–818.

Gordon, P. & Richardson. H. (1997). Are compact cities a desirable planning goal? *Journal of the American Planning Association*, 63(1), 95–106.

Gore, E.H. (1951). *From Florida sand to The City Beautiful*. Winter Park, FL: Orange Press.

Gou, J.Y. & Bhat, C.R. (2007). Operationalizing the concept of neighborhood: Application to residential location choice analysis. *Journal of Transport Geography*, 15, 31–45.

Govers, R. (2011). From place marketing to place branding and back. *Place Branding and Public Diplomacy*, 7(4), 227–231.

Greenberg, M.R. (1999). Improving neighborhood quality: A hierarchy of needs. *Housing Policy Debate*, 10(3), 601–624.

Hackworth, J. & Rekers, J. (2005). Ethnic packaging and gentrification: The case of four neighborhoods in Toronto. *Urban Affairs Review*, 41(2), 211–236.

Haeberle, S.H. (1987). Neighborhood identity and citizen participation. *Administration & Society*, 19(2), 178–196.

Handy, S.L., Xing, Y. & Buehler, T.J. (2010). Factors associated with bicycle ownership and use: A study of six small U.S. cities. *Transportation*, 37(6), 967–985.

Hankins, K.B. (2007). The final frontier: Charter schools as new community institutions of gentrification. *Urban Geography*, 28(2), 113–128.

Hayes, K.J. & Taylor, L.R. (1996). Neighborhood school characteristics: What signal quality to homebuyers? *Federal Reserve Bank of Dallas Economic Review*, Fourth Quarter, 2–9.

Hidalgo, M.C. & Hernandez, B. (2001). Place attachment: Conceptual and empirical questions. *Journal of Environmental Psychology*, 21(3), 273–281.

Hudak, S. (2018). Orange county history center's Pulse exhibit stirs memories, hearts, and tears. Retrieved from: www.orlandosentinel.com/news/orange/os-history-center-pulse-exhibit-20180605-story.html

I-4 Ultimate (2018). Project overview. Retrieved from: https://i4ultimate.com/project-info/overview/

Instagram (2018). @baldwinparkbubble. Retrieved from: www.instagram.com/baldwinparkbubble/

Jacobs, J. (1961). *The death and life of great American cities*. New York: Vintage.

Johansson, O. & Cornbiese, M. (2010). Place branding goes to the neighbourhood: The case of pseudo-Swedish Andersonville. *Geografiska Annaler: Series B, Human Geography*, 92(3), 187–204.

Kaczynski, A.T., Potwarka, L.R. & Sealens, B.E. (2008). Association of park size, distance, and features with physical activity in neighborhood parks. *American Journal of Public Health*, 98(8), 1451–1456.

Kardan, O. et al. (2015). Neighborhood greenspace and health in a large urban center. *Scientific Reports*, 5, article 11610.

Kasadra, J.D. & Janowitz, M. (1974). Community attachment in mass society. *American Sociological Review*, 39(3), 328–339.

Keatinge, B. & Martin. D.G. (2016). A "Bedford Falls" kind of place: Neighbourhood branding and commercial revitalization in processes of gentrification in Toronto, Ontario. *Urban Studies*, 53(5), 867–883.

Kelling, G.L. & Wilson, J.Q. (1982). Broken windows. Retrieved from: www.theatlantic.com/magazine/archive/1982/03/broken-windows/304465/

Kelsh, J. (2015). Neighborhood branding and marketing. Retrieved from: www.neighborworks.org/Documents/Community_Docs/Revitalization_Docs/StableCommunities_Docs/Defining-the-New-Brand.aspx

Ketter, E. & Avraham, E. (2012). The social revolution of place marketing: The growing power of users in social media campaigns. *Place Branding and Public Diplomacy*, 8(4), 285–294.

Knez, I. (2005). Attachment and identity as related to a place and its perceived climate. *Journal of Environmental Psychology*, 25(2), 207–218.

LaGrange, T.C. (1999). The impact of neighborhoods, schools, and malls on the spatial distribution of property damage. *Journal of Research in Crime and Delinquency*, 36(4), 393–422.

LaGrone, P. (2017). I-4 named the most dangerous highway in America. Retrieved from: www.abcactionnews.com/news/local-news/i-4-named-the-most-dangerous-highway-in-america

Lake Nona (2018). About us. Retrieved from: www.lakenona.com/about/

Lewicka, M. (2011). Place attachment: How far have we come in the last 40 years? *Journal of Environmental Psychology*, 31(3), 207–230.

LIFT Orlando (2018). LIFT Orlando. Retrieved from: www.liftorlando.org

Lotan, G.T. (2017). Federal report on Pulse: Authorities performed well but more training needed. Retrieved from: www.orlandosentinel.com/news/pulse-orlando-nightclub-shooting/os-pulse-shooting-after-action-report-20170707-story.html

Ma, L. & Dill, J. (2017). Do people's perceptions of neighborhood bikeability match "reality"? *Journal of Transport and Land Use*, 10(1), 291–308.

Maas, J. et al. (2006). Green space, urbanity, and health: How strong is the relation? *Journal of Epidemiological Community Health*, 60, 587–592.

Mannarini, T. & Fedi, A. (2009). Multiple senses of community: The experience and meaning of community. *Journal of Community Psychology*, 37(2), 211–227.

Manzo, L.C. & Perkins, D.D. (2006). Finding common ground: The importance of place and attachment to community participation and planning. *Journal of Planning Literature*, 20(4), 335–350.

Martin, A. (2017). Parents worried as Orange schools prepare to close. Retrieved from: www.orlandosentinel.com/features/education/school-zone/os-orange-school-closings-20170201-story.html

Masuda, J.R. & Bookman, S. (2018). Neighbourhood branding and the right to the city. *Progress in Human Geography*, 42(2), 165–182.

McMillan, D.W. & Chavis, D.M. (1986). Sense of community: Definition and theory. *Journal of Community Psychology*, 14(1), 6–23.

Meltzer, M. (2017). 12 neighborhoods across America that are about to get crazy popular. Retrieved from: www.thrillist.com/travel/nation/the-next-big-neighborhoods-in-america-new-york-la-miami-chicago

MetroWest Master Association (2018). MetroWest community. Retrieved from: www.metrowestcommunity.com

Mills 50 (2018). Retrieved from: www.mills50.org

Morrow-Jones, H.A, Irwin, E.G. & Roe, B. (2004). Consumer preference for neotraditional neighborhood characteristics. *Housing Policy Debate*, 15(1), 171–202.

Nall, C. (2015). The political consequences of spatial policies: How interstate highways facilitated geographic polarization. *Journal of Politics*, 77(2), 394–406.

Nandan, S. (2004). An exploration of the brand identity–brand image linkage: A communications perspective. *Brand Management*, 12(4), 264–278.

National Association of Realtors (2018). Walter W. Rose. Retrieved from: www.nar.realtor/walter-w-rose

No author (1913). Orlando, Florida. No publisher identified.

One Orlando Alliance (2018). Act, love, give. Retrieved from: https://oneorlandoalliance.org/acts-of-love-and-kindness/

Orlando Economic Partnership (2018). You don't know the half of it. Retrieved from: www.orlandoedc.com/Why-Orlando/You-Dont-Know-the-Half-of-It.aspx

Pais, J. et al. (2014). Neighborhood reputation and resident sentiment in the wake of the Las Vegas foreclosure crisis. *Sociological Perspectives*, 57(3), 343–363.

Pedicini, S. (2017). Visit Orlando: Record 68 million people visited last year. Retrieved from: https://www.orlandosentinel.com/business/tourism/os-visit-orlando-tourist-numbers-20170511-story.html.

Perkins, D.D. & Long, D.A. (2002). Neighborhood sense of community and social capital: A multi-level analysis. In: Fisher, A., Sonn, C., & Bishop, B. (eds) *Psychological sense of community: Research, applications, and implications*. New York: Plenum, pp. 291–318.

Peroldo, R. (2017). The great freezes and the collapse of the Florida citrus industry. Retrieved from: https://medium.com/florida-history/the-great-freezes-1894-95-and-the-collapse-of-the-florida-orange-industry-7442e5d75337

Phillips, B.J., McQuarrie, E.F. & Griffin, W.G. (2014). The face of the brand: How art directors understand visual brand identity. *Journal of Advertising*, 43(4), 318–332.

Pilkington, E. (2010). How the Disney dream died in celebration. Retrieved from: www.theguardian.com/world/2010/dec/13/celebration-death-of-a-dreamPostal, L. (2017). New school for gifted students to open in Orange next year. Retrieved from: www.orlandosentinel.com/features/education/school-zone/os-gifted-school-elementary-students-orange20171026-story.html

Postal, L. (2018). New school for gifted named Orlando gifted academy. Retrieved from: www.orlandosentinel.com/features/education/school-zone/os-orlando-gifted-academy-new-school-20180529-story.html

Powell, L.M. et al. (2006). Availability of physical activity–related facilities and neighborhood demographic and socioeconomic characteristics: A national study. *American Journal of Public Health*, 96(9), 1676–1690.

Redmon, K.C. (2010). The man who reinvented the city. Retrieved from: www.theatlantic.com/personal/archive/2010/05/the-man-who-reinvented-the-city/56853/

Reedy Creek Improvement District (2018). About us. Retrieved from: www.rcid.org/about/

Retting, R. (2018). Pedestrian traffic fatalities by state. Retrieved from: www.ghsa.org/sites/default/files/2018-03/pedestrians_18.pdf

Rich, M.A. & Tsistos, W. (2016). Avoiding the "SoHo effect" in Baltimore: Neighborhood revitalization and arts and entertainment districts. *International Journal of Urban and Regional Research*, 40(4), 736–756.

Roe, J.J., Aspinall, P.A. & Ward Thompson, C. (2017). Coping with stress in deprived urban neighborhoods: What is the role of green space according to life stage? *Frontiers in Psychology*, 8, doi: 10.3389/fpsyg.2017.01760

Russon, G. (2018). 72 million tourists visited Orlando in 2017, a record number. Retrieved from: www.orlandosentinel.com/business/tourism/os-bz-visit-orlando-tourism-2017-story.html

Sampson, R.J., Raudenbush, S.W. & Earls, F. (1997). Neighborhoods and violent crime: A multilevel study of collective efficacy. *Science*, 277, 918–924.

Schuler, G. (2018). Florida is the second worst state in the country for pedestrian deaths, says study. Retrieved from: www.orlandoweekly.com/Blogs/archives/2018/03/06/florida-is-the-second-worst-state-in-the-country-for-pedestrian-deaths-says-study

Schultz, E.J. (2017). Inside the ad agency helping Vegas mourn – and plot its comeback. Retrieved from: https://adage.com/article/cmo-strategy/agency-helping-vegas-mourn-plotting-comeback/310826/

Sentinel Print (1910). *Orlando: The charm of the south*. Orlando, FL: Sentinel Print.

Sevin, E. (2013). Places going viral: Twitter usage patterns in destination marketing and place branding. *Journal of Place Management and Development*, 6(3), 227–239.

Shelton, N. (2018). Orlando sets another U.S. travel record with 72 million visitors. Retrieved from: www.visitorlando.com/blog/index.cfm/2018/5/11/Orlando-Sets-Another-US-Travel-Record-With-72-Million-Visitors/

Silver, E. & Miller, L.L. (2004). Sources of informal social control in Chicago neighborhoods. *Criminology*, 42(3), 551–584.

Singh, G.K., Siahpush, M. & Kogan, M.D. (2010). Neighborhood socioeconomic conditions, built environments, and childhood obesity. *Health Affairs*, 29(3), 503–512.

Siordia, C. & Saenz, J. (2013). What is a "neighborhood"? Definition in studies about depressive symptoms in older adults. *Journal of Frailty and Aging*, 2(3), 153–164.

Smiley, K.T., Rushing, W. & Scott, M. (2016). Behind a bicycling boom: Governance, cultural change and place character in Memphis, Tennessee. *Urban Studies*, 53(1), 193–209.

Stark, R. (1987). Deviant places: A theory of the ecology of crime. *Criminology*, 25(4), 893–910.Sugiyama, T. (2012). The built environment and physical activity behavior change: New directions for research? *Research in Exercise Epidemiology*, 14(2), 118–124.

Sugiyama, T. et al. (2013). Initiating and maintaining recreational walking: A longitudinal study on the influence of neighborhood green space. *Preventative Medicine*, 57, 178–182.

Sun, T. et al. (2006). Online word-of-mouth (or mouse): An exploration of its antecedents and consequences. *Journal of Computer-Mediated Communication*, 11(4), 1104–1127.

Symon. E. (2017). 6 bizarre realities of living in a town owned by Disney. Retrieved from: www.cracked.com/personal-experiences-2442-so-perfect-its-creepy-i-live-in-city-designed-by-disney.html

Talen, E., Menozzi, S. & Schaefer, C. (2015). What is a "Great Neighborhood?" *Journal of the American Planning Association*, 81(2), 121–140.

Taylor, R.B. (1996). Neighborhood responses to disorder and local attachments: The systemic model of attachment, social disorganization, and neighborhood use value. *Sociological Forum*, 11(1), 41–74.

The National Main Street Council (2018). About us. Retrieved from: www.mainstreet.org/about-us

Tiebout, C. (1956). A pure theory of local expenditures. *Journal of Political Economy*, 64(5), 416–424.

Twigger-Ross, C.L. & Uzzell, D.L. (1996). Place and identity processes. *Journal of Environmental Psychology*, 16(3), 20–220.

Tyson, C.J. (2014). Municipal identity as property. *Penn State Law Review*, 118(3), 647–696.

Ujang, N. (2012). Place attachment and continuity of urban place identity. *Procedia – Social and Behavioral Sciences*, 49, 156–167.U.S. Census Bureau (2018). Census tracts. Retrieved from: www.census.gov/geo/reference/webatlas/tracts.html

Vignoles, V.L. et al. (2006). Beyond self-esteem: Influence of multiple motives on identity construction. *Journal of Personality and Psychology*, 90(2), 308–333.

WFTV (2014). Community fights to keep fern creek elementary from closing. Retrieved from: www.wftv.com/news/local/community-meeting-planned-over-closing-fern-creek-/107393593

Wherry, F.W. (2011). *The Philadelphia barrio: The arts, branding, and neighborhood transformation.* Chicago, IL: University of Chicago Press.

White, A.W. (1965). *History of development in Orange and Seminole counties: Growth patterns of urban form in the Orlando metropolitan area.* Orlando, FL: No publisher identified.

Willits, D., Broidy, L. & Denman, K. (2013). Schools, neighborhood risk factors, and crime. *Crime & Delinquency,* 59(2), 292–315.

Younan, D. et al. (2016). Environmental determinants of aggression in adolescents: Role of urban neighborhood greenspace. *Journal of the American Academy of Child Adolescent Psychology,* 55(7), 591–601.

Index